Gravitation and Modern Cosmology
The Cosmological Constant Problem

**ETTORE MAJORANA
INTERNATIONAL SCIENCE SERIES**
Series Editor:
Antonino Zichichi
European Physical Society
Geneva, Switzerland

(PHYSICAL SCIENCES)

Recent volumes in the series:

Volume 49   **NONLINEAR OPTICS AND OPTICAL COMPUTING**
Edited by S. Martellucci and A. N. Chester

Volume 50   **HIGGS PARTICLE(S): Physics Issues and
Experimental Searches in High-Energy Collisions**
Edited by A. Ali

Volume 51   **BIOELECTROCHEMISTRY III: Charge Separation
Across Biomembranes**
Edited by G. Milazzo and M. Blank

Volume 52   **ELECTROMAGNETIC CASCADE AND CHEMISTRY OF
EXOTIC ATOMS**
Edited by Leopold M. Simons, Dezso Horvath, and Gabriele Torelli

Volume 53   **NEW TECHNIQUES FOR FUTURE ACCELERATORS III:
High-Intensity Storage Rings—Status and Prospects
for Superconducting Magnets**
Edited by Gabriele Torelli

Volume 54   **OPTOELECTRONICS FOR ENVIRONMENTAL SCIENCE**
Edited by S. Martellucci and A. N. Chester

Volume 55   **SEMIMAGNETIC SEMICONDUCTORS AND DILUTED
MAGNETIC SEMICONDUCTORS**
Edited by Michel Averous and Minko Balkanski

Volume 56   **GRAVITATION AND MODERN COSMOLOGY:
The Cosmological Constant Problem**
Edited by Antonino Zichichi, Venzo de Sabbata,
and Norma Sánchez

Volume 57   **NEW TECHNOLOGIES FOR SUPERCOLLIDERS**
Edited by Luisa Cifarelli and Thomas Ypsilantis

Volume 58   **MEDIUM-ENERGY ANTIPROTONS AND THE
QUARK–GLUON STRUCTURE OF HADRONS**
Edited by R. Landua, J.-M. Richard, and R. Klapisch

A Continuation Order Plan is available for this series. A continuation order will bring delivery of each new volume immediately upon publication. Volumes are billed only upon actual shipment. For further information please contact the publisher.

Gravitation and Modern Cosmology
The Cosmological Constant Problem

Volume in honor of
Peter Gabriel Bergmann's
75th birthday

Edited by

Antonino Zichichi
European Physical Society
Geneva, Switzerland

Venzo de Sabbata
Università di Bologna
Bologna, Italy

and

Norma Sánchez
Observatoire de Paris–Meudon
Meudon, France

Plenum Press • New York and London

Library of Congress Cataloging-in-Publication Data

Gravitation and modern cosmology : the cosmological constant problem /
  edited by Antonino Zichichi, Venzo de Sabbata, and Norma Sánchez.
      p.   cm. -- (Ettore Majorana international science series.
  Physical sciences ; v. 56.)
    "Proceedings of a symposium held in honor of Peter Gabriel
  Bergmann's 75th birthday, held September 17-20, 1990, in Erice,
  Sicily, Italy"--T.p. verso.
    Includes bibliographical references and index.
    ISBN 0-306-44054-7
    1. Gravitation--Congresses.   I. Zichichi, Antonino. II. de
  Sabbata, Venzo.   III. Sanchez, N. (Norma), 1952-   . IV. Bergmann,
  Peter Gabriel. V. Series.
  QC178.G63 1991
  530.1'4--dc20   91-40369
                                                                    CIP

Proceedings of a symposium held in honor of Peter Gabriel Bergmann's
75th birthday, held September 17–20, 1990, in Erice, Sicily, Italy

ISBN 0-306-44054-7

© 1991 Plenum Press, New York
A Division of Plenum Publishing Corporation
233 Spring Street, New York, N.Y. 10013

All rights reserved

No part of this book may be reproduced, stored in a retrieval system, or transmitted
in any form or by any means, electronic, mechanical, photocopying, microfilming,
recording, or otherwise, without written permission from the Publisher

Printed in the United States of America

Bergmann and Einstein working in Einstein's study at Princeton in early 1938, when they were actively engaged in attempts at constructing unitary field theories. (Photo by Lotte Jacobi.)

P.G. Bergmann and A. Zichichi at Erice in September 1990.

PREFACE

Peter Gabriel Bergmann started his work on general relativity in 1936 when he moved from Prague to the Institute for Advanced Study in Princeton. Bergmann collaborated with Einstein in an attempt to provide a geometrical unified field theory of gravitation and electromagnetism. Within this program they wrote two articles together: A. Einstein and P.G. Bergmann, Ann. Math. 39, 685 (1938) ; and A. Einstein, V. Bargmann and P.G. Bergmann, Th. von Karman Anniversary Volume 212 (1941). The search for such a theory was intense in the ten years following the birth of general relativity. In recent years, some of the geometrical ideas proposed in these publications have proved essential in contemporary attempts towards the unification of all interactions including gravity, Kaluza-Klein type theories and supergravity theories.

In 1942, Bergmann published the book "Introduction to the Theory of Relativity" which included a foreword by Albert Einstein. This book is a reference for the subject, either as a textbook for classroom use or for individual study. A second corrected and enlarged edition of the book was published in 1976. Einstein said in his foreword to the first edition: "Bergmann's book seems to me to satisfy a definite need... Much effort has gone into making this book logically and pedagogically satisfactory and Bergmann has spent many hours with me which were devoted to this end."

As is well known in classical electrodynamics, the Maxwell equations of the electromagnetic field do not determine the equations of motion of the sources which are given independently. On the contrary, in general relativity, the Einstein equations governing the dynamics of the gravitational field also determine the equations of motion of the sources. The demonstration of this fundamental property is far from being trivial; in particular because of the singularities of the field at the points where particles are located, and the corresponding self-energy. The proof was pioneered by Einstein, Infeld and Hoffmann (in 1938) for point masses, and later improved in a number of subsequent papers. The same result was independently obtained by Fock (in 1939) for extended sources, involving a series of approximations. Bergmann was able to derive the equations of motion in an invariant and elegant manner, independently of the choice of coordinates and of the approximation procedure, using general principles like conservation laws, covariance, and the equivalence of mass and energy. This had important consequences for the computation of gravitational radiation from accelerated gravitating systems and the post-Newtonian effects in celestial mechanics.

From the early fifties onwards, one of the main goals and challenging tasks for Bergmann was the quantization of the gravitational field. The difficulties which he faced and solved were enormous in dealing with the full non-linear properties of Einstein theory and with non-perturbative approaches. Bergmann and his collaborators investigated the canonical (i.e., Hamiltonian) formulation of classical and quantum covariant field theories. Bergmann formulated and studied the difficult problem of the *constraints* which arise from the existence of the coordinate-transformation group as an invariance group of the theory (not all the conjugate momenta, nor all the field equations themselves, are dynamically independent). Bergmann analyzed the group theoretical significance of Dirac's Hamiltonian formulation (and the generalized Poisson bracket in three dimensions), prepared the grounds for its quantum description, and clarified and made profound contributions in the different quantization programs based on the canonical formalism. As Bergmann himself said, "the resulting theory would give us answers to such questions as the nature of a fully quantized geometry of space-time, the role of world points in this geometry, the 'softening-up of the light cone', and the effect of this not only on the divergences associated with the gravitational field but with all other fields as well".

We will not describe all of Bergmann's work here. We would however like to mention the great contribution that Bergmann made to the School of Cosmology and Gravitation in Erice from its inception in 1974: being ever present, always discussing every argument in great depth, and always clarifying every aspect of the problems under discussion. During the Sixth Course of the School (1979) he discussed "the fading world point" dealing with the nature of space-time and of its elements. During the Eighth Course (1983), he discussed various unitary theories such as Kaluza-Klein, scalar-tensor theories and projective theories. During the Ninth Course (1985) devoted to "Topological Properties of Space-Time" and during the Tenth Course (1987) devoted to "Gravitational Measurements", Bergmann discussed "gravitation at spatial infinity" and the "observables in general relativity", where he showed in a very elegant way the profound difference between the notion of observable in general relativity and the corresponding concept in special relativity (or in Newtonian physics).

A Symposium in honor of Peter Gabriel Bergmann's 75th Birthday, devoted to the discussion of the cosmological constant problem, took place in Erice from 17-20 September 1990. Current topics in cosmology and gravitation were discussed and this book gathers together contributions presented at the Symposium, together with invited papers submitted by other colleagues who could not be present at the Symposium but who nevertheless wished to honor Peter Gabriel Bergmann.

The reader will find in this volume an updated version of different approaches to the cosmological constant problem, as well as contributions on classical and quantum cosmology, cosmic and quantum strings, classical general relativity, and gravitational radiation and its experimental search.

We wish to express our grateful thanks to all the authors: some have contributed review papers, and others reports on their recent research work.

We thank the staff members of the Ettore Majorana Centre at Erice for their valuable help both in the organization of the Symposium and in the different stages of realization of this book.

We admire Peter Gabriel Bergmann both as a scientist and as a man of great culture. He represents for the younger generations an example to be followed. With these feelings we have devoted some of our time to pay tribute to a great physicist whose life has been entirely dedicated to the progress of Science with its genuine values.

Antonino Zichichi
Venzo de Sabbata
Norma Sánchez

CONTENTS

My Life 1
    Peter G. Bergmann

Effective Action Model for the Cosmological
    Constant Revisited 5
    Stephen L. Adler

Could Final States of Stellar Evolution Proceed Towards
    Naked Singularities? 11
    Nicola Dallaporta

Torsion, Quantum Effects and the Problem of
    Cosmological Constant 19
    Venzo de Sabbata and C. Sivaram

Variations of Constants and Exact Solutions in
    Multidimensional Gravity 37
    S.B. Fadeev, V.D. Ivashchuk and V.N. Melnikov

Larger Scale Structure in The Lyman-Alpha
    Absorption Lines 51
    Li-Zhi Fang

Null Surface Canonical Formalism 59
    J. N. Goldberg, D.C. Robinson and C. Soteriou

Qualitative Cosmology 65
    I.M. Khalatnikov

Third Quantization of Gravity and the Cosmological
    Constant Problem.. 87
  G. Lavrelashvili, V. A. Rubakov and P. G. Tinyakov

Cosmological Constant, Quantum Cosmology
    and Anthropic Principle... 101
  A. Linde

On the Gravitational Field of an Arbitrary
    Axisymmetric Mass with a Magnetic Dipole Moment...................... 121
  Igor D. Novikov and Vladimir S. Manko

Twistors as Spin 3/2 Charges.. 129
  Roger Penrose

Experimental Search of Gravitational Waves... 139
  Guido Pizzella

A Simple Model Of the Universe without Singularities........................... 151
  Nathan Rosen and Mark Israelit

String Theory and the Quantization of Gravity....................................... 157
  N. Sánchez

Projective Unified Field Theory in Context with
    the Cosmological Term and the Variability
    of the Gravitational Constant... 179
  Ernst Schmutzer

The Introduction of the Cosmological Constant...................................... 185
  E.L. Schucking

Pre-Post-History of Tolman's Cosmos.. 189
  E.L. Schucking, S. Lauro and J.-Z. Wang

Some Ideas on the Cosmological Constant Problem................................ 201
  G. Veneziano

Velocity of Propagation of Gravitational Radiation, Mass of the Graviton, Range of the Gravitational Force, and the Cosmological Constant.. 217
   J. Weber

Contributors... 225

Index... 227

MY LIFE

Peter G. Bergmann

Departments of Physics
Syracuse University
Syracuse, NY 13244-1130
New York University
New York, NY 10003

First of all, let me express my deep appreciation to all of you who have contributed to this get-together, to Professors A. Zichichi and V. de Sabbata, to Dr. A. Gabriele and the other staff members of the Ettore Majorana Centre, to my colleagues and friends, those who contributed talks and those who came to be part of a joyous community, and also to those who were unable to join us but who sent greetings. Indeed, all of us may congratulate each other that we have been privileged to take part in an endeavor that satisfies us intellectually and that contributes, we are confident, to a better understanding of the physical universe.

I became a university student in 1931, beginning at the Technical University (then "Hochschule") at Dresden, where I registered both for chemistry and physics laboratory exercises. I found these laboratory courses thoroughly enjoyable, even though I was committed to majoring eventually in theoretical physics. Most undergraduates in the United States hate their freshman physics laboratories. Having taught laboratory sections at Lehigh University myself, I am not surprised. At Dresden, Professor Harry Dember had organized a first-year laboratory with over one-hundred different experiments. For each experiment there was only one set-up available, and thus only two students could perform that experiment together, in one afternoon. There were experiments in mechanics, in heat, in optics, and so forth. I recall that one afternoon we examined the output of a photoelectric cell as a function of intensity and wavelength of the light source.

Obviously, the experiments could not be synchronized with the subjects dealt with in the demonstration lectures, as each pair of students had to be scheduled differently. But each of us gained some intuitive feeling for the relative accuracies that could be achieved with optical devices, with an analytic balance, with calorimetric techniques, and others. I found it exciting to be able to determine myself the heat of fusion of ice, or the value of the gravitational constant g near sea level. - I might add that the equipment was well maintained, and that each set-up had a permanent location, from which it was never removed.

In the United States the laboratory exercises are usually synchronized with the big lectures, so that all students in the course perform the same experiment the same week, requiring numerous copies of the same experiment, which must be set up, and removed, to make room for the next scheduled experiment. This arrangement carries with it a certain degree of rigidity, aside from the much greater capital investment required. There is no denying the advantage of having the laboratory exercises illustrate the ideas propounded in the lecture, but I, for one, preferred the informality and the greater variety offered in Professor Dember's arrangement.

Having spent two semesters in Dresden, and two more at Freiburg, I intended to go for the remainder of my student years to the University of Berlin. I understood that there were many outstanding theoretical physicists at Berlin, and presumably many opportunities to begin research leading to the doctorate. But by the spring of 1933 Hitler had become the leader of the German government, the Reichstag building had been torched, and being a Jew I was unable to register as a student. In the fall of 1933 I became a student at the German University of Prague, one of three academic institutions operated by the Czechoslavak Republic for those of its citizens whose native tongue was German.

I had selected Prague because I did not need to learn a new language of instruction, and it was far less expensive than, for instance, the Swiss universities. Once I was there, I found the professor of theoretical physics, Philipp Frank, to be deeply interested in the foundations of physics, and an excellent teacher if one wanted to understand the motivations underlying physical theories. I have never shared Frank's positivistic attitudes, - he was a leading member of the so-called Vienna circle, - but under his tolerant guidance I developed a taste and a concern for the epistemological and ontological underpinnings of theoretical physics.

In 1936 I received my doctorate degree. I had written to Albert Einstein, telling him of my eagerness to work under him, and had received an encouraging response. When I called on him in the fall of that year, he was working with L. Infeld and B. Hoffmann on the theory of motion of ponderable bodies immersed in a gravitational field, now known as the EIH theory. General relativity differs from other (classical) field theories in that the laws of motion of the sources of the gravitational field are a consequence of the field equations. By contrast, in electrodynamics the law of motion of charges, the Lorentz force, is independent of Maxwell's equations, which govern the behavior of the electromagnetic field. One can move the electric charges on arbitrary trajectories; there are still solutions of Maxwell's equations compatible with these sources. Not so with gravitating masses: If they are to be moved along predetermined trajectories, one requires tools to move them, and these tools, in turn, affect the gravitational field in such a manner that the sources and the field together satisfy the field equations.

As for me, Einstein first suggested that I attempt to construct a combination of a gravitational and electromagnetic field of a kind that might represent an electron. When we were able to prove that such a model did not exist, Einstein returned to the search for a unitary field theory, that is to say, a generalization of his general theory of relativity that would combine all known physical interactions into a single field. Together with Valya Bargmann, who came to Princeton in 1937, one year after myself, we looked at theories based on Kaluza's higher-dimensional proposal, at Weyl's conformal geometry, and at a number of other possibilities, all of them highly speculative, and, ultimately, none of them truly satisfactory.

Einstein's sustained search for a unitary field theory is to be explained, I think, by his philosophic views, which differed significantly from that of most contemporary physicists. Quantum theory associates with any physical system a complex structure, variously known as the Schrödinger wave function or the Dirac ket. This ket represents fully the state of the system. But knowledge of the ket does not automatically fix the values of all observable quantities associated with the system. Rather, knowledge of the quantum state, that is of the ket, merely asserts the probabilities of the possible outcomes of measurements. Moreover, if a measurement is actually performed, this manipulation changes the ket, so that all probabilities need to be recalculated. As the ket itself is not an experimentally observable physical quantity, according to quantum theory, physical reality resides in the relationship between the physical system and the observer. That the physical world about us appears to exist whether we choose to observe it or to leave it alone, is to be explained by the large size of most of the objects surrounding us. The same holds for the causality that appears to govern the behavior of large objects. At the atomic and the subatomic level neither the outcome of measurements nor the evolution in the course of time are fully predictable.

On the other hand, quantum theory explains why a number of physical variables can assume only discrete values. Foremost among these are the angular momentum of an isolated physical system and its total energy, at least for bound states. The great triumphs of quantum theory have been the successful analyses of the energy levels of atoms and molecules, and subsequently of atomic nuclei and of subatomic systems.

Though he was one of the pioneers of early quantum physics, Einstein insisted on the independent existence of the objects outside ourselves, with all their physical properties. He felt equally strongly about the inherent causality governing natural phenomena. "The Lord does not play dice" was one of his favorite sayings. Throughout his later life, Einstein was convinced that eventually quantum phenomena would be recognized as resulting from the very involved interactions of a classical (non-quantum) physical field, that did not require the ministrations of a human observer. Of course, he did not deny that the intervention of an observer, equipped with measuring devices, would affect the object observed.

After Einstein's death in 1955 the search for a unitary field theory fell in disfavor, for it was recognized that without more detailed knowledge of the nature of nuclear forces this search was mostly speculation. A quest for a theoretical structure based solely on one's subjective view of logical simplicity did not appear promising. In recent years, given a much more detailed knowledge of subatomic particles and their interactions, many physicists have resumed their search for a theory that "explains everything", but usually with quantum physics built right into the foundations.

For me Einstein has been a wonderful teacher. Though he was conscious of the value of his contributions to physics, he never used his stature as an argument in a scientific discussion. He worked closely with a few younger colleagues, but he made no attempt to attract a large group of people to form a "school". Einstein was in his late fifties when I first met him. Being a true master of his native tongue, German, he keenly felt the lack of equal ease in talking and writing in English. To the end of his lfe he drafted his scientific papers in German, which someone else translated into English. Much of this translating was done by Valya and Sonya Bargmann.

I stayed in Princeton for five years. Then it was time for me to find an academic teaching position. The Second World War called on most of us to exchange the academic environment for military research. After the war, in 1947, I joined the physics faculty at Syracuse University, an association that I treasure to this day. Teaching at the graduate level led me to broaden my interests in areas that were not closely related to relativity. Perhaps more important, contacts with graduate students, with postdoctorate associates, and with fellow faculty have developed into friendships that gave me at least as much as I contributed.

My principal research interest has been shaped by my association with Einstein. Whether one agrees with Einstein's philosophy or prefers the "mainstream" views, there can be little question that general relativity and quantum theory are this century's most important conceptual developments in physics. Whereas quantum theory revolutionized our conceptions of the relationship between subject and object, general relativity has provided a totally new view of the structure of space and time. Are these two conceptual frameworks compatible with each other? If not, does one or the other require major modification so as to provide physics with a single all-encompassing conceptual basis? I do not believe that we know the answers to this day. To attempt to quantize the gravitational field means to approach these questions with technical means. Both eventual success and eventual failure should provide valuable insights, and this hope has motivated much of my work.

In 1955 Wolfgang Pauli organized the first international conference on general relativity, which met at Berne. For the first time in my life I met most of the senior people in my field. Since that time work in general relativity and gravitation has grown tremendously. Whereas at Berne fewer than one hundred colleagues met, a typical conference today attracts well over one thousand participants. There are at least two scientific journals devoted exclusively to general relativity. Many more papers are being published in general purpose journals. What is perhaps more important than quantity, there is increasing diversity. Today research in general relativity includes large experimental programs, the design of gravitational wave antennas being the most conspicuous. Cosmology and cosmogony, and the discovery and investigation of exotic astronomical objects, all are intimately tied together. At the subatomic level we inquire into the meaning of space and time intervals of the order of Planck dimensions, that is to say, of hypothetical particles whose Schwarzschild radii (proportional to the mass) and Compton wavelengths (inversely proportional to the mass) are equal. These investigations naturally involve ties to high-energy physics.

In my lifetime general relativity has changed from an abstract subject cultivated by a few devotees to a well integrated and popular part of mainstream contemporary physics. What it has lost in intimacy it has gained in relevance. We have every right to hope that the generations of relativists that follow us will contribute much to our understanding of nature.

EFFECTIVE ACTION MODEL FOR THE COSMOLOGICAL CONSTANT REVISITED

Stephen L. Adler

Institute for Advanced Study
Princeton, NJ 08540, USA

ABSTRACT

I give a revised version of an effective action model for the vanishing of the cosmological constant, incorporating the observation of Y. H. Gao that it is better to formulate the argument with a Mellin transform than with a Laplace transform.

INTRODUCTION

Two years ago I wrote a short paper[1] formulating Coleman's[2] Euclidean space argument for the vanishing of the cosmological constant (see also Banks[3], Baum[4] and Hawking[5]) from an effective action point of view. Subsequently, Y. H. Gao[6] pointed out that my argument, which was formulated using a Laplace transform containing an unspecified lower limit of integration, can be formulated without arbitrary limits of integration by using a Mellin transform. Since this observation appears not to have been published, I thought it would be useful to use the occasion of this volume, in honor of Peter G. Bergmann's 75th Birthday, to revisit the effective action model for the vanishing of the cosmological constant, taking Gao's suggested use of a Mellin transform into account.

As in my original argument, I start from the Euclidean world partition function

$$Z = \int d[g_{\mu\nu}]d[\phi]e^{-S[g_{\mu\nu},\phi]}, \qquad (1)$$

with $g_{\mu\nu}$ the metric and ϕ the matter (all non-metric) degrees of freedom. Let us introduce a general coordinate covariant regularization prescription[7] characterized by a cutoff M, with $M \ll M_P$ (the Planck mass), and use it to factorize the integral into an integration over degrees of freedom with length $\ell \geq M^{-1}$, and a second integration over all shorter length scales,

$$Z = \int_{\ell \geq M^{-1}} d[g_{\mu\nu}]d[\phi]\, e^{-S^M_{eff}[g_{\mu\nu},\phi]},$$

$$e^{-S^M_{eff}[g_{\mu\nu},\phi]} \equiv \int_{\ell \leq M^{-1}} d[g_{\mu\nu}]d[\phi]\, e^{-S[g_{\mu\nu},\phi]}. \qquad (2)$$

Gravitation and Modern Cosmology
Edited by A. Zichichi *et al.*, Plenum Press, New York, 1991

The standard quantum gravity analysis[8] invokes general covariance and locality to write for $S_{eff}^M[g_{\mu\nu}, \phi]$ the general form

$$S_{eff}^M[g_{\mu\nu}, \phi] = -\frac{1}{16\pi G_M} \int (R - 2\Lambda_M)(g)^{1/2} d^4x$$
$$+ \text{ higher powers of } R + \text{matter action}, \qquad (3)$$

with the lowest two powers in a local curvature expansion indicated explicitly. The new ingredient added in Refs. 2 and 3 is the observation that since S_{eff}^M is an effective action it is *not* expected to be a local functional, but can have nonlocal terms as well, arising in particular from fluctuations in spacetime topology on the Planck length scale. The simplest nonlocality (a sum over wormhole "handles" on one closed subuniverse, assuming the Hartle-Hawking boundary condition) adds to Eq. (3) a term of the form

$$\left[\int (g)^{1/2} d^4x \right]^2, \qquad (4)$$

but it is clear that in general higher powers of $\int (g)^{1/2} d^4x$ can appear as well.[9] As an effective action embodiment of Coleman's model, I will adopt the Ansatz that

$$S_{eff}^M[g_{\mu\nu}, \phi] = -\frac{1}{16\pi G} \int R(g)^{1/2} d^4x$$
$$+ F_M\left(\int (g)^{1/2} d^4x \right) + \ldots, \qquad (5)$$

with $F_M(V)$ a *nonlinear* function of its argument V; I have also assumed that renormalizations of the gravitational coupling coming from length scales greater than M^{-1} are insignificant, so that G_M can be replaced by Newton's constant G (which is related to the Planck mass by $G = M_P^{-2}$).

I will now show two things: (i) The postulated form of the effective action in Eq. (5) is preserved when we change the cutoff mass M to a lower mass M', in other words, our Ansatz scales into the infrared in a consistent fashion, and (ii) the Ansatz of Eq. (5), for generic F_M, implies vanishing of the fully renormalized cosmological constant.

To show (i), we note that by definition

$$e^{-S_{eff}^{M'}[g_{\mu\nu},\phi]} = \int_{(M')^{-1} \geq \ell \geq M^{-1}} d[g_{\mu\nu}] d[\phi] \, e^{-S_{eff}^M[g_{\mu\nu},\phi]}. \qquad (6)$$

Following Gao[6], we now write $\exp(-F_M)$ as a Mellin transform

$$e^{-F_M\left(\int (g)^{1/2} d^4x\right)} = \int_0^\infty \frac{d\sigma}{\sigma} f_M(\sigma) \, \sigma^{-\int (g)^{1/2} d^4x}$$

$$= \int_0^\infty \frac{d\sigma}{\sigma} f_M(\sigma) \, e^{-(\ell n \sigma) \int (g)^{1/2} d^4 x} , \tag{7}$$

so that the right-hand side of Eq.(6) becomes

$$\int_0^\infty \frac{d\sigma}{\sigma} f_M(\sigma) \int_{(M')^{-1} \geq \ell \geq M^{-1}} d[g_{\mu\nu}] d[\phi]$$

$$\times \exp \left\{ \frac{1}{16\pi G} \int [R - 2\Lambda_M(\sigma)] (g)^{1/2} d^4 x + \ldots \right\} , \tag{8a}$$

with

$$\Lambda_M(\sigma) = 8\pi G \ell n \sigma \tag{8b}$$

ranging from $-\infty$ to ∞ as σ ranges from 0 to ∞. The inner integral in Eq. (8a) involves just a standard gravitational action (with the cosmological constant Λ_M a function of the parameter σ). Hence neglecting topological fluctuations (which like renormalizations of the gravitational coupling are not expected to be significant on length scales much larger than the Planck length) the inner integral becomes

$$\int_{(M')^{-1} \geq \ell \geq M^{-1}} d[g_{\mu\nu}] d[\phi] \exp \left\{ \frac{1}{16\pi G} \int [R - 2\Lambda_M(\sigma)] (g)^{1/2} d^4 x + \ldots \right\}$$

$$= \exp \left\{ \frac{1}{16\pi G} \int \left[R - 2\Lambda_{M'}\left(\Lambda_M(\sigma)\right) \right] (g)^{1/2} d^4 x + \ldots \right\} , \tag{9}$$

with the effective cosmological constant $\Lambda_{M'}$ at length scales $(M')^{-1}$ some function of $\Lambda_M(\sigma)$. Defining a new transform variable σ' by

$$\Lambda_{M'}\left(\Lambda_M(\sigma)\right) = 8\pi G \ell n \sigma' ,$$

$$\frac{d\sigma}{\sigma} f_M(\sigma) \equiv \frac{d\sigma'}{\sigma'} f_{M'}(\sigma') , \tag{10}$$

Eqs. (6)-(9) combine to give

$$S_{eff}^{M'}[g_{\mu\nu}, \phi] = -\frac{1}{16\pi G} \int R(g)^{1/2} d^4 x + F_{M'}\left(\int (g)^{1/2} d^4 x\right) , \tag{11a}$$

with

$$e^{-F_{M'}(\int (g)^{1/2} d^4 x)} = \int_0^\infty \frac{d\sigma'}{\sigma'} f_{M'}(\sigma')(\sigma')^{-\int (g)^{1/2} d^4 x} . \tag{11b}$$

This completes the demonstration that the Ansatz of Eq. (5) is invariant in form when we change the cutoff from M to a smaller mass M'. Note that in general

$F_{M'}(V)$ is not the same function of its argument as $F_M(V)$. [An exception to this statement occurs when $F_M(V)$ is linear in V, as in the conventional gravitational action, so that $f_M(\sigma)$ is proportional to a delta function; in this case $f_{M'}(\sigma)$ will also be proportional to a delta function and $F_{M'}(V)$ will again be linear in V. Clearly, as long as the change of variable of Eq. (10) is smooth (as is expected), a smooth $f_M(\sigma)$ cannot evolve as M is changed to a delta function, or equivalently, a generic nonlinear $F_M(V)$ cannot evolve to a linear $F_{M'}(V)$.]

To demonstrate (ii), let us substitute the Ansatz of Eq. (5), together with the Mellin transform representation of Eq. (7), into the integral for the full partition function in Eq. (2), giving

$$Z = \int_0^\infty \frac{d\sigma}{\sigma} f_M(\sigma) \int_{l \geq M^{-1}} d[g_{\mu\nu}] d[\phi]$$
$$\times \exp\left\{\frac{1}{16\pi G} \int [R - 2\Lambda_M(\sigma)](g)^{1/2} d^4x + \ldots \right\} . \quad (12)$$

Since the inner integral is a standard gravitational partition function (with the intermediate cosmological constant Λ_M dependent on a parameter σ), the usual analysis[10] gives

$$\int_{l \geq M^{-1}} d[g_{\mu\nu}] d[\phi] \exp\left\{\frac{1}{16\pi G} \int [R - 2\Lambda_M(\sigma)](g)^{1/2} d^4x + \ldots \right\} = e^{-\Gamma_\sigma(\bar{g}_{\mu\nu}, \bar{\phi})} , \quad (13)$$

with the background fields $\bar{g}_{\mu\nu}$ and $\bar{\phi}$ the ones which minimize the functional Γ_σ, requiring

$$\delta_{g_{\mu\nu}} \Gamma_\sigma(g_{\mu\nu}, \phi)\bigg|_{\bar{g}_{\mu\nu}, \bar{\phi}} = \delta_\phi \Gamma_\sigma(g_{\mu\nu}, \phi)\bigg|_{\bar{g}_{\mu\nu}, \bar{\phi}} = 0 . \quad (14)$$

Making a curvature expansion of $\Gamma_\sigma(g_{\mu\nu}, \phi)$ gives

$$\Gamma_\sigma(g_{\mu\nu}, \phi) = -\frac{1}{16\pi G} \int [R - 2\Lambda(\sigma)](g)^{1/2} d^4x + \ldots , \quad (15)$$

with $\Lambda(\sigma)$ the fully renormalized cosmological constant for given σ. Changing from σ to $\Lambda(\sigma)$ as integration variable in the outer integral by writing

$$\frac{d\sigma}{\sigma} f_M(\sigma) \equiv d\mu(\Lambda) , \quad (16)$$

we can rewrite the partition function as

$$Z = \int d\mu(\Lambda) \, e^{-\Gamma_\Lambda(\bar{g}_{\mu\nu}, \bar{\phi})} , \quad (17a)$$

with the Λ integration extending from $-\infty$ to ∞, where we have defined

$$\Gamma_\Lambda(g_{\mu\nu}, \phi) \equiv \Gamma_{\sigma(\Lambda)}(g_{\mu\nu}, \phi) = -\frac{1}{16\pi G} \int [R - 2\Lambda](g)^{1/2} d^4x + \ldots . \quad (17b)$$

Equation (17) thus has the form of an ensemble of gravitational actions with varying cosmological constant, integrated with measure $d\mu(\Lambda)$. As long as $d\mu(\Lambda)$ is smooth and nonvanishing at $\Lambda = 0$ [which will be true for generic nonlinear $F_M(V)$ but *not* for $F_M(V)$ linear in V], we can follow Hawking's argument for the evaluation of Eq. (17): We neglect the influence of real matter ϕ, so that Eq. (14) becomes

$$\delta_{g_{\mu\nu}} \Gamma_\Lambda(g_{\mu\nu}, 0)\Big|_{\bar{g}_{\mu\nu}} = 0 , \qquad (18)$$

the solutions of which are Einstein spaces (solutions of the source-free Einstein equations with cosmological constant). For negative Λ, $\Gamma_\Lambda(\bar{g}_{\mu\nu}, 0)$ is positive, while for positive Λ, the four-sphere of radius $(3\Lambda^{-1})^{1/2}$ is the metric $\bar{g}_{\mu\nu}$ which minimizes $\Gamma_\Lambda(g_{\mu\nu}, 0)$ and the action at the minimum is negative,

$$\Gamma_\Lambda(\bar{g}_{\mu\nu}, 0) = -\frac{3\pi}{\Lambda G} . \qquad (19)$$

Substituting Eq. (19) into Eq. (17a) gives a non-integrable essential singularity at $\Lambda = 0$,

$$Z = \int d\mu(\Lambda) \, e^{3\pi/\Lambda G} , \qquad (20)$$

and hence $\Lambda = 0$ completely dominates all other contributions to the integral over Λ in Eq. (17a). Thus the observed cosmological constant is zero.

To summarize: the vanishing of the cosmological constant follows, by a straightforward application of functional integral methods, from the Ansatz of Eq. (5). All that is required is that the intermediate effective action S^M_{eff} be a nonlinear function of the space-time volume V; the nature of the nonlinearity, and the energy scale M used to separate "Planck scale" from "larger than Planck scale" gravitational physics, are not essential to the analysis. The argument given here is clearly close in spirit to that of Ref. 2, and in fact the one-subuniverse approximation to Coleman's derivation, in which the $\Lambda = 0$ contribution to Z is exponentially enhanced as in Eq. (20), corresponds to Eqs. (5) and (7) above with $f_M(\sigma)$ chosen to be a Gaussian in $\ln \sigma$. Coleman's full wormhole calculus, in which the sum over multiple closed subuniverses leads to an enhancement factor of $\exp\exp(3\pi/\Lambda G)$ in his analog of Eq. (20), can also be expressed as an effective action argument, but requires the inclusion of curvature-dependent terms in the non-linear part F_M. This elaboration, however, is not necessary to capture the essence of the argument for the vanishing of Λ, and the sum over multiple subuniverses is what in fact leads to certain difficulties [11] with Coleman's original calculation.[12] The virtue of the approach given here is that it shows clearly that the vanishing of the cosmological constant is a property of a wide class of intermediate effective actions, and is not tied to specific (e.g., dilute-gas) approximations. The Ansatz of Eq. (4) may also serve as a bridge to understanding the connection between Euclidean and Minkowski space arguments for the vanishing of Λ, perhaps through the application of the closed-time-path functional formalism.[13]

I wish to thank Ed Witten, Natti Seiberg and Greg Moore for a stimulating conversation which prompted the original version of this investigation. I am grateful to

Howard Georgi for calling my attention to Y. H. Gao's preprint, and to the American Physical Society for permission to use a modified version of Ref. 1 for this volume. This work was supported by the U.S. Department of Energy under grant number DE-FG02-90ER40542. To conclude, I wish to extend very best birthday greetings to Professor Peter G. Bergmann and his family.

References

1. S. L. Adler, Phys. Rev. Lett. **62**, 373 (1989).
2. S. Coleman, Nucl. Phys. **B310**, 643 (1988).
3. T. Banks, Nucl. Phys. **B309**, 493 (1988).
4. E. Baum, Phys. Lett. **133B**, 185 (1984).
5. S. W. Hawking, Phys. Lett. **134B**, 403 (1984).
6. Yi Hong Gao, "A Note on Adler's Argument for the Vanishing of the Cosmological Constant", unpublished preprint from the Department of Modern Physics, University of Science and Technology of China, Hefei, Anhui, P. R. China (1988).
7. The fact that general relativity is the zero slope limit of string theory shows that such regularizations do, in principle, exist.
8. For a survey, see "General Relativity," S.W. Hawking and W. Israel, eds. (Cambridge University Press, Cambridge, 1979).
9. This is explicitly noted by Banks in Ref. 3, and was first emphasized to me by N. Seiberg (private communication).
10. For a detailed exposition, see S.L. Adler, Rev. Mod. Phys. **54**, 729 (1982), Sec. VI.
11. See e.g., J. Polchinski, Phys. Lett. **B219**, 251 (1989).
12. However, it is not completely clear whether Coleman's $\exp\exp(3\pi/\Lambda G)$ enhancement is a general result, or is specific to his dilute-gas approximation. It requires a *coherence* between the effective actions in the multiple closed universes, which could be washed out by corrections to the dilute-gas approximation. For instance, if the effective Λ's in the different subuniverses differ by random "noise," the $\exp\exp(3\pi/\Lambda G)$ is smeared into a single exponential.
13. See e.g., E. Calzetta and B.L. Hu, Phys. Rev. **D35**, 495 (1987).

COULD FINAL STATES OF STELLAR EVOLUTION PROCEED TOWARDS NAKED SINGULARITIES?

Nicola Dallaporta

Dipartimento di Astronomia, Università di Padova
Vicolo Osservatorio, 535623 Padova, Italia

Having the privilege of being present during this manifestation aiming to honour Professor Bergmann on the occasion of his 75th birthday, I am of course wishing to participate in expressing my admiration for his several scientific achievements, although I am not a relativist my self. Thus, being not in the position of offering him some personal contribution in his domain, I will present in this meeting the work on an interesting issue of General Relativity done by Professor Fernando de Felice, who has been once my scholar: he has of course allowed me to do so, and is glad on his own behalf for this opportunity of expressing indirectly his own appreciation on Professor Bergmann's work.

The problem considered by de Felice and his occasional collaborators is as follows. The discovery of pulsars and their successful interpretation as rotating neutron stars has generally been considered as a proof for gravitational collapse occurring in nature as predicted by General Relativity. However, the theory of superdense matter allows neutron stars to form only if their mass does not exceed a critical value M_c of the order of 3 $M_\odot$; for larger values, hard core repulsion of nuclei would not be in stand any more to hinder further collapse, and the expected final state should lead to the formation of a black hole. Now, while for average main sequence stars, stellar evolution foresees a collapsing core below this critical mass M_c, giving thus rise to the observed normal pulsars, it is expected that upper main sequence stars, with masses, let us say, above 12-15 $M_\odot$ should give rise to a core value exceeding M_c, which therefore should collapse into a black hole; and it is believed that, in some few cases concerning close binaries, experimental data are best interpreted by assuming that the collapsed component of the pair is indeed a black hole.

However, the possibility for the final collapsed state being a black hole is submitted to the condition (Misner et al.[1] (1973)) that:

$$a^2 + Q^2 \leq m^2 \qquad (1)$$

where $m = GM/c^2$, $a = J/Mc$, $Q = G^{1/2} q/c^2$
and M being the total mass, J the total angular momentum, q the total electric charge of the collapsed object, G the gravitational constant and c the velocity of the light.
As the total charge of celestial bodies is most likely zero, the preceding triction (1) practically reduces to:

$$a^2 \leq m^2 \quad \text{so that} \quad a/m = cJ/GM^2 \leq 1$$

So, should a/m exceed one, the Kerr solution of the Einstein equations should predict the occurrence of a naked singularity.

Now, it is quite possible for all main sequence stars for which the relevant data are known, to calculate the ratio a/m, provided some suitable assumption is made concerning the type of rotation. de Felice and coworkers [2] have considered two different prescriptions: A) rigid body rotation, for stars with masses ranging between 1 and 10 $M_\odot$, and equatorial velocities of the order of 200~400 Km/sec, obtaining values for a/m of the order of 10 to 100. B) as such situation is not very likely to occur owing to the fact that convective motions in the stellar interior are expected to redistribute angular momentum, they have also used several models of massive stars (15- 300 $M_\odot$) constructed by Bodenheimer [3], with various angular momentum distribution laws. The results of the combined A) and B) cases are that 1 $M_\odot$ stellar cores have a ratio a/m ranging in the same interval of values (18- 100).

Instead, when applying the same calculations for neutron stars, with standard values for radius $R \sim 10^6$ cm and mass $M \sim 1$ $M_\odot$ and constant density throughout, assumed to rotate rigidly, they find even for the fastest rotating case associated with the Crab nebula, with period P= 33. 10^{-3} sec., an a/m value of 1,7 . 10^{-2}, that is quite small with respect to unity.

All this means of course that for middle main sequence stars some phenomena occur during the intermediate evolutionary phases, contributing to the decrease of the a/m ratio in the core by 3 to 4 orders of magnitude. Could we then expect the same should occur for upper main sequence stars, in order that the a/m ratio of their core should be diminished below one, and thus give rise to the expected black hole?

The problem considered by de Felice can then been summarized in the following two steps : 1) it is possible to identify and understand processess which must be present during stellar evolution and specially during the final collapsing phase, to decrease the a/m value in the required amount as observed in neutron stars? 2) Once identified, should these same processes
be responsible for a similar decrease of a/m in larger mass stars ending thus possibly into black holes?

I will now briefly review the different phases of de Felice's work, and conclude that up to now the situation is far from being well cut, so that the problem remains for the moment widely open, and leaves without answer the question of what we should consider as being the ultimate state of collapse for large mass stars, should we fail in identifying appropriate phenomena occurring for lowering the a/m value below one.

As a first logical step, one may look what changes could be brought to a/m during the evolution from the main sequence to the red giant stage with double shell burnings; in such a transition, the helium rich core of the star is strongly contracting with a corresponding expansion of the surrounding enveloppe, so that angular momentum will be transferred from core to envelope. Models worked out by Endal and Sofia [4] with realistic redistribution of angular momentum, transferred mainly by convection and Eddington circulation, have been used, and it was found on the whole that the a/m ratio during post main sequence evolution decreases by an amount of no more than 40 50 %, so the range of possible values extends now between 5 and 50, which is still too high in respect to neutron stars. However, the possible effects of the magnetic field have not been included in those models, and the opinions concerning their importance are rather widespread.

The difficuly in much lowering the a/m ratio during the red giant phase leaves no other possibility than looking for effects during the formation stage of the neutron star, that is during or immediatly after core collapse. a) As a first point, an obvious correction is likely due to the well known fact that the presence of an off-centered dipole magnetic field in the neutron star provoking continuous radiation acts as a decelerating factor on its rotational energy ; this has in fact been observed in many pulsars and provokes a lengthening of their rotation period. The well known formula for such a period increase is given by:

$$\dot{P} = \frac{1}{2} \frac{P}{t} \left(1 - \frac{P_0^2}{P^2}\right) \qquad (2)$$

where P_0 is the initial rotation period, P its value after time t, $\dot{P}$ its variation in unit time. P, $\dot{P}$, and t being well known for the Crab pulsar, one gets : $P_0 = 17.10^{-3}$ sec., a still larger value than $P_1 = 0,57.10^{-3}$ sec. corresponding to a/m = 1. Therefore even its initial a/m value immediately after collapse was smaller than one. For other pulsars we do not know exactly their age t; however, their rotation is generally rather longer, so according to eq. (2) one easily foresees that their initial P_o should correspond to an a/m value smaller than one. Therefore the consideration of radiation deceleration appears not to be sufficient in leading to an initial a/m value compatible with the red giant stage models.

b) A more promising result follows from the consideration that most pulsars appear to be high velocity objects with translational velocities whose average probable lower limit is of the order of 120 km/ sec. This can be explained, according to Harrison and Tademaru [5], as due to some inherent anisotropy in the supernova outburst, leading to the emission of asymmetric electromagnetic radiation, the reaction to which being a translational acceleration of the collapsing core at the expense of its initial rotational energy. The formula for such an effect given by these authors reads:

$$MV_o = \frac{\varepsilon}{2c} I (\Omega_L^2 - \Omega_o^2)$$

where I is the momentum of inertia of the pulsar, V_o its final translational energy, Ω_L the initial angular velocity of the collapsing core, Ω_o that of the neutron star at the end of the accelerating period and ε the radiation asymmetry coefficient. For the Crab pulsar with Ω_o = 3. 7 10^2 sec^{-1} corresponding to its P_o value, V_o = 120 km/ sec , I = 10^{45} gram cm^2 and $\varepsilon \sim 10^{-2}$, one gets Ω_L = 1,28. 10^4 sec^{-1} , corresponding to an initial period of P_o = 0, 49. 10^{-3} sec implying a/m $\sim$ 1.

So, the following deductions can now be expressed:

I) Provided the Harrison-Tademaru mechanism should take place, as we could reach at the end of the red giant stage an a/m value of the order of 5-50, one should have to look for a further mechanism allowing to fill the still existing gap between 5-50 > a/m > 1;

II) Should instead the Harrison -Tademaru mechanism, whose only experimentl support is the observed average translational velocity of pulsars, not be of a general occurrence, or should the asymmetry coefficient ε be still smaller than here asummed, then the gap in the a/m value between the final stage of stellar evolution and the neutron star stage should remain as wide as about three orders of magnitude.

III) If we now wish to extrapolate the discussion also to large mass stars, for which we expect the collapsed core to turn into a black hole, we have to look for a solution analogous to the case II), as the further decrease of a/m in the case of a neutron star due to the Harrison-Tademaru effect is achieved in a time scale much longer than the dynamical time of collapse, which has only to be considered in the black hole case . So we are compelled in requiring a decrease of a/m from the order 5-50 to a value just smaller than unity, which is the minimum request for obtaining a black hole as the ending stage for the core collapse.

In all the three cases , the crucial phase to be considered is the collapse itself, the transition between the core in the ending pre-supernova stage and the final collapsed state after the outburst. Two main phenomena have been considered and analyzed by de Felice and collaborators : 1) mass shedding due to centrifugal forces during collapse; 2) emission of gravitational radiation during the same phase.

1) Let us consider first , the mass shedding: the hint for looking at such phenomenon starts from computer calculations due to Nakamura [6] and Nakamura and Sato [7] concerning gravitational collapse for a rotating mass . These authors find that when a/m is greater than one , there is a tendency for the outer material towards having its infall halted and of part of it remaining as a bound ring around the central mass which continues to collapse . These numerical calculations for some specific cases have been generalized by considering a gravitational collapse not halted by pressure of a rotating asymmetric fluid configuration, the condition imposed on which for mass shedding as taking place being the usual condition of balance between gravitational and centrifugal forces.

In a first paper [8] it was assumed that no redistribution of specific angular momentum is occurring during collapse, so J and M are constant, and therefore the values of a/m is assumed to be conserved until shedding starts. It is found that the centrifugal condition:

$$a/m = \kappa \ F \ r_e / r_g$$

where r_e is the equatorial radius when shedding starts, and r_g is the Schwarzschild radius-, is of course also dependent on κ, the form factor of the momentum of inertia, and on F, a factor depending on the shape of the configuration. After shedding has begun, one finds that the variation of a/m is given by:

$$\delta(a/m) = (\delta m)(h - 2a)$$

where h = J/ KM is the specific angular momentum, so a/m decreases only if h > 2 a. Several cases are discussed , but on the whole the decrease of a/m, going along with the decrease of the central mass due to shedding, should never exceed some 40% of its initial value.

In a further paper [9], a more general case has been considered , with redistribution of angular momentum before and during shedding: the discussion is somewhat complicated and cannot be quickly summarized ; it is perhaps enough saying that generally speaking , the existence and the increase of an outer structure due to equatorial shedding forming a ring is not a guarantee that the a/m value of the remaining inner structure should be bound to necessarily decrease ; so that on the whole shedding phenomenon might turn out in some cases as being very inefficient for preparing the suitable conditions for the formation of a black hole.Moreover, as mass shedding by itself substracts matter from the central collapsing core , it may act in the sense of reducing its mass below the critical value M_c, so that a neutron star and not a black hole should be the outcome of the collapse.

2) Looking now for the second effect related to post newtonian quadrupole gravitational wave emission, the general case of a triaxial ellipsoidal structure (with $a_1 > a_2 > a_3$) undergoing gravitational collapse has been considered [10]. The final results appear to depend on three main parameters: the initial value of the a/m ratio ; the ratio of the two equatorial axes and the qualitative behaviour of the a/m ratio for the collapsing core as a function of the ratio of r to the Schwarzschild radius is practically insensitive to contraction as long as r/r_g is larger than few unities , and decreases substantially towards one only in the very last phases when r_g is approached by r. This reduction however is sensible only if the initial value of a/m is significantly greater than one , otherwise the variation is unconspicuous. On the whole, whatever the initial a/m , it does not appear than one could reach in the final situation values much lower than one.

As a last step , in order to examine the whole extension of the possible variatons of the a/m value, de Felice and coworkers have moreover considered: A) the evolution of a/m during star formation from an original cloud, preceding the main sequence stage of stars [1], which is of course widely dependent on dissipative processes largely unknown. They find that the dynamics of collapse of initially extended mass clouds, -whose a/m value is generally very large ($\sim 10^5$)- evolving towards the presequence stage, is mostly determined by the ratio of the rotational to the gravitational energy. The process is generally dominated by successive fragmentation phases of the original cloud; although the fragments mostly carry a small fraction of the mass and angular momentum of the initial cloud, their a/m ratio does not necessarily decrease.The need of a substantial decrease of the a/m ratio of the fragments gives to this quantity a role of guarantee of internal consistency to any numerical model of star formation. Up to now, the uncertainties present in the whole process do not allow following the path along which a/m should decrease from the high initial values to the smaller ones corresponding to stars.

B)A further interesting consideration refers to the formation of binary systems from the rotating interstellar clouds [12], as neutron stars and eventually black hole formation is generally believed to take place mostly in such systems. An accurate analysis for binaries for which all necessary data are available reveals that, although the largest amount of angular momentum of the initial cloud is taken by the orbital motion, with a/m of the order of some hundreds, its values for the spin of both components, assuming a complete phase locking state, is of the same order as for single stars. Some theoretical models of the process due to Bodenheimer [13] allow to reproduce rather successfully this order of magnitude observed, while others fail to produce a net losse of the a/m ratio in fragments, casting doubts about some evolution towards values consistent with the ones observed in normal stars. More recent calculations however, indicate that protostellar contraction, under the effects of tidal interaction with nearby objects of comparable masses, assures a net decrease (one order of magnitude) of the a/m ratio in all models considered. On the whole ther is no indication that binaries might be responsible for the low a/m of pulsars.

Finally, in a recent paper [14], de Felice and collaborators have turned considering the formation of large black holes in the center of spiral galaxies, as is suggested by the analysis of the rotational data very near to the center of galaxies such as M31. With an adequate fluid model for the bulge, they look for the consistency between the derived data and the existence of a massive black hole in the very central part of it, based on the requirement that the value of a/m should not exceed unity.A rather detailed discussion indicates that the rotation velocities of the nucleus of M31 are not consistentent with the presence of a giant black hole with mass of the order of $10^7 M_\odot$ in the center of the galaxy. However, the uncertainties connected with the structure of the galaxies brings back to the problem of understanding the mechanisms which efficiently decrease the a/m ratio.In the case of galaxies the difficulty of this problem is much larger than in the case of stars, and the variety of phenomena which would occur during the collapse of the core do not allow to draw very definite conclusions in this case.

The ensemble of the results here exposed seem to point out that we are still somewhat far from understanding how black holes do form as final states of the evolution of the central parts of stars and till now,of galaxies.de Felice conclusions can be summarized as follows:

1) the ratio a/m does not seem to change in a significant way during the whole course of stellar evolution, although some efficient mechanism, such as the Harrison-Tademaru effect-, should be operating either during the collapse to neutron star, or soon after the neutron star formation, in order to explain the low a/m value observed for neutron stars.

2) the considered effects for reduction of a/m over dynamical time scales of collapse are mass shedding and gravitational radiation; in the former case however,the reduction of mass due to the shedding tends to favoring the formation of neutron stars rather than black holes; in the second case, the effect occurs only at very late stages of collapse; and if the collapse is sufficiently fast, and not leading to a neutron star, it is not clear wether gravitational radiation is able to lowering always a/m below one, as the cosmic censorship principle would require. However, this might not be considered as an a priori drawback, as presently there are already several examples of space-times violating this assumption; so the difficulty of lowering the a/m ratio below one,

as put into evidence by the work of de Felice, is now beginning to be considered as an indication for the possible occurrence of naked singularity solutions as ending states following the final collapse of evolved stars; and specific models for such solutions are now being considered and studied   [15].

REFERENCES

1. P. Misner, K. Thorne, A. Wheeler, "Gravitation", Freeman , N.Y.(1973).
2. F. de Felice and Yu Yunqiang, J. Phys. A 15, 3341 (1982).
3. P. Bodenheimer, Ap.J. 167, 153 (1971).
4. A.S. Endal and S. Sophia, Ap.J. 210, 148 (1976); Phys. Rev. Lett.39,1429 (1977); Ap.J. 220, 279 (1978); 232, 539 (1979).
5. F.R.Harrison and E. Tademaru, Ap.J. 201, 447 (1975).
6. T. Nakamura, Progr. Theor. Phys. 65, 187 (1981).
7. T.Nakamura and H. Sato, Progr.Theor.Phys. 66, 2038 (1981).
8. J.C. Miller and F. de Felice, Ap.J.298,474 (1985).
9. F. de Felice and Yu Yunqiang, Month. Not. 220,737 (1986).
10. F. de Felice, J.C. Miller and Yu Yunqiang, Ap.J.298,480 (1985).
11. F. de Felice , Mem. Soc. Astr. Ital.54 (1983).
12. F. de Felice and L. Sigalotti, Proc.XIV Yamada Conference, Kyoto, World Sc.Pub. (1986).M. N. R. A.S., 1991,in press.
13. P. Bodenheimer, Ap.J.224, 488 (1978).
14. M. Bradley, A. Curir and F. de Felice, Univ.of Torino report, (1981).
15. N. Charlton and C.J.S. Clarke, Class. Q. Grav. 7 , 743 (1990).

TORSION, QUANTUM EFFECTS AND THE PROBLEM

OF COSMOLOGICAL CONSTANT

Venzo de Sabbata and C.Sivaram

World Laboratory - Lausanne
Department of Physics of Ferrara University
Ferrara, Italy Indian Institute of Astrophysics
Bangalore, India

1. INTRODUCTION

This symposium on the cosmological constant problem is being held concurrently with the workshop dealing with challenge in Tev physics. If one considers that the highest conceivable energy scale in either accelerator physics or cosmology is the Planck energy $(\hbar c^5/G)^{1/2} \simeq 10^{19}$ Gev = E_{Pl} and noting that the energy of the 2.7° Kelvin cosmic microwave background photon is $E_\gamma \simeq 10^{-13}$ Gev (in high energy parlance!) then we have the interesting coincidence (to be taken in lighter spirit!):

$$(E_{PL} \times E_\gamma) \simeq \text{one Tev!} \qquad (1)$$

Thus the Tev scale emerging as the geometric mean between the highest and lowest energy scales in big bang cosmology may serve as a unifying link connecting the two symposia on seemingly very different topics! After all the big bang was the biggest accelerator one could ever have had, but unfortunately now is in the deceleration phase with very low energy of 10^{-13} Gev! Of course we also have the postulated baryon number violation at Tev scales with its own profound cosmological consequences including relic Tev particles! Since this has been dealt with by other people we would not talk about this aspect any further in this article!

The so called cosmological constant Λ was omitted by Einstein in his first formulation of the field equation that

Gravitation and Modern Cosmology
Edited by A. Zichichi *et al.*, Plenum Press, New York, 1991

he derived starting from Poisson equation $\Delta\phi = 4\pi G\rho$ and generalizing it in order that results in covariant form and substituting to the gravity force the curved space-time through the metric tensor $g_{\mu\nu}$. In this way Einstein is led to postulate the equation:

$$G_{\mu\nu} = \chi T_{\mu\nu} \qquad (2)$$

where $G_{\mu\nu} = R_{\mu\nu} - (1/2)g_{\mu\nu}R$. But when Einstein like to apply his equations to cosmology, he was forced to introduce a term $\Lambda g_{\mu\nu}$: in fact Einstein was looking for a static, homogeneous and isotropic universe and these features were not compatible with his equations; at this point Einstein had the good intuition that in order to balance the global gravitational attraction one has to introduce a term that in some way acts as a repulsive cosmological force. As this term was introduced for the purely cosmological purpose, Λ had the name of 'cosmological constant'. The equation (2) then will be

$$R_{\mu\nu} - (1/2)g_{\mu\nu}R - \Lambda g_{\mu\nu} = \chi T_{\mu\nu} \qquad (3)$$

At that time Einstein believed that the field equations (3) had no possible solution for empty space (i.e. for $\rho = 0$). But when de Sitter was able to find a solution also for $\rho = 0$ Einstein was seriously disturbed mainly because this is incompatible with the Mach principle understood as 'inertia fully and esclusively determined by matter' and from that moment he became against the Λ - term (today we know that de Sitter solution is the basis for most works in inflation). When later on Hubble discovered his law showing the expansion of the universe, Einstein, referring also to the Friedman theoretical work "which was not influenced by experimental facts", rejected definitely the Λ - term and declared that the introduction of the cosmological term was his greater mistake. However once the genie of the cosmical constant was let out of the bottle it was difficult to bury it or cap it down and in fact was rejuvenated several times in different contexts to serve different cosmological purposes [1]!

The fact that a cosmological term can give an increased Hubble age has often been invoked whenever there is a contradiction between the age of the universe end the age of its oldest constituents usually the globular clusters containing the oldest Population II stars. Recent observations such as the Th-232/Nd data [2], and from quasar double images [3] imply that we must have a greater Hubble age (present estimates being too low as the more recent for the Hubble constant H_0 which imply a lower age [4]) which in turn requires the presence of a Λ-term.

But the strongest argument in favour of the cosmological term comes from quantum field theory: in fact it was realized that the vacuum energy momentum tensor in quantum field theory must be uniquely of the form $(T_{\mu\nu})_{vac} = \rho_{vac} g_{\mu\nu}$, i.e. anything that contributes to the energy density of the vacuum acts just like a cosmological constant, the effective vacuum energy density being of the form:

$$\rho_{vac} = \Lambda_{eff} c^4/8\pi G \qquad (4)$$

Now there would have been large changes in the vacuum energy in the early universe as a result of phase transitions due to the breaking of some symmetry group as the expanding universe cooled. If the symmetry breaking takes place at some energy M, then the induced vacuum energy density is $\approx M/(\hbar/Mc)^3 \propto M^4$. For instance at the Planck epoch when $t \approx 10^{-43}$ sec and $M \approx 10^{19}$ Gev one would expect a large vacuum energy density term $\rho_{vac} \approx 10^{114}$ ergs cm$^{-3} \approx 10^{78}$ Gev fm^{-3} $(= E_{Pl} l_{Pl}^{-3} = c^7/G^2\hbar)$ where 1 fm = 10^{-13} cm, corresponding to an effective cosmological constant of $\Lambda_{Pl} \approx 10^{66}$ cm^{-2}, at this epoch. Λ_{Pl} could arise as a result of scale invariance breaking and quantum gravitational contributions to the vacuum energy in the very early universe at the Planck epoch [5,6]. The GUT phase transition at energies $\approx 10^{15}$ Gev, would similarly induce another large Λ of $\Lambda_{GUT} \approx 10^{50}$ cm^{-2}. There would be also other large contributions from other symmetry breaking phase transitions at the electroweak scale for instance (at energies $\approx 10^2$ Gev, $\Lambda \approx 10^{-3}$ cm^{-2}) and perhaps at the Tev scale. These happen at later epochs in the early universe. The question is what has happened to all these large contributions to the Λ-term, Why is the present value of Λ so small? Observations involving galaxy clusters, etc. enables us to estimate the Hubble constant H_0 and consequently the critical density for a closed universe as

$$\rho_c = 3H_0^2/8\pi G \simeq 10^{-44} \text{Gev fm}^{-3} \quad (10^{-29} \text{g/cm}^3) \quad (5)$$

implying a present maximal value for Λ (on the basis that the vacuum energy density cannot much exceed the critical density) as $\Lambda_{obs(max)} \leq 10^{-56} \text{cm}^{-2}$. This would imply a ratio of

$$\Lambda_{obs}/\Lambda_{Pl} \simeq \rho_c/\rho_{Pl} = 3H_0^2 G^2 \hbar/8\pi Gc^5 = 3\hbar G H_0^2/8\pi c^5 \simeq 10^{-122} \quad (6)$$

which would make it the smallest dimensionless number in physics. However it must be borne in mind that even such a small value of the limit on the observed cosmological constant Λ_{obs} implies a vacuum density exceeding the visible matter density by a factor of nearly a hundred! This by itself would have significant consequences for cosmology and astrophysics as such a residual Λ term can increase the age of the universe to agree with estimates of oldest stellar populations even with a higher value of H_0 [7] and/or higher density of darke matter. We shall return to the question of the residual value of the cosmological constant at the present epoch, later. The point is that in flat space estimates, the energy of the vacuum even if indefinitely large cancels out of the physical processes, i.e. in all estimates of transition probabilities between two states, what is relevant is the energy difference between two states, the transition is from vacuum + state 1 to vacuum + state 2. This is not true once gravitational fields are present, as the gravitational field couples to the actual amount of energy present rather than to an energy difference, i.e. without gravity one does not know how much the 'absolute zero' of energy is. So the cosmological constant problem is what has happened to all the large vacuum energies present in the early universe since the gravitational dynamics of the present universe imply a very low if not zero value of the residual vacuum energy as given by the Λ-term (vastly less by 120 orders of magnitude by what one expects from particle physics).

2. SOLUTIONS PROPOSED FOR THE COSMOLOGICAL CONSTANT PROBLEM

Weinberg [8], while giving a good current review of the situation, summarizes five different approaches undertaken in recent years to understand this question.

He considers supersymmetry (exact global supersymmetry would indeed make the vacuum energy, and hence Λ, vanish since the fermion and boson contributions cancel). But we know that supersymmetry must be broken quite strongly at the low energies in the present universe (if it was ever present at all) and this would give a large contribution to Λ which would not vanish. Λ_{Pl} would still survive! Unlike gauge invariance in electromagnetism implying zero-photon mass there is no symmetry principle known which would make Λ vanish exactly. There is no specific property or principle in supergravity or superstring theory which is known to make Λ vanish or very small! Furthermore string theories with non-zero cosmological constants cannot represent true ground states or vacua of a string field theory because of a non-zero dilaton correlation function and such theories are thus not even well defined. The vast majority of string theories do not even have the property of equal numbers of bosonic and fermionic degree of freedom. Weinberg also resorts to the anthropic principle (as a last resort?) to explain the smallness of the effective cosmological constant.

The anthropic bound follows from the requirement that the Hubble time must be comparable to the lifetime of main-sequence stars (which evolve life)[9] given by $t_s \simeq \alpha^2 \alpha_G^{-1}(m_p/m_e)^2 \hbar/m_p c^2$ (α = fine structure constant, $\alpha_G = (Gm_p/\hbar c)$, m_p, m_e, are the proton and electron masses. This fixes Λ as:

$$\Lambda/\Lambda_{Pl} \leq \alpha^{-4}(m_e/m_p)^2(m_p/m_{Pl})^6 \qquad (7)$$

A somewhat similar but however non-anthropic bound on Λ by imposing the operational constraint that for the vacuum energy shifts over atomic or nuclear scales to be physically meaningfull they should be at least measurable over Hubble time scales, gives[10]:

$$\Lambda \simeq 6\hbar H_0 m_e^3 c^2 G/e^6 \approx 10^{-57} cm^{-2} \qquad (8)$$

(e is the electron electric charge).
Other attempts to understand the cosmological constant problem have involved some sort of adjustment mechanism [11,12] which usually requires some extra scalar field, which evolves and acts as a counterterm to cancel the cosmological term. However it turns out that in all such attempts the scalar field must

have some very special ad hoc properties and involves a lot of 'fine-tuning' at all stages; apart from there being no evidence of such extra fields. Other endeavours have dealt with changing the structure of Einstein's equations [8] but this also leads to several consistency problems.

Recently there has been a lot of excitement about a new mechanism suggested by Coleman [13] which follows up an earlier work of Hawking which described how in quantum cosmology there could arise a distribution of values for the effective cosmological constant with an enormous peak at $\Lambda_{eff} = 0$. Coleman considers the effects of topological fixtures known as wormholes, consisting of two asymptotically flat spaces joined together at a 3-surface which may be a tube or throat. A classical example is the Einstein-Rosen bridge [14] which is a slice joining two asymptotic regions. Coleman shows that the probability distribution or expectation values has an infinite peak at $\Lambda_{eff} \Rightarrow 0$. However several objections to this approach have been raised: (i) the reality of the existence of wormholes, (ii) the use of Euclidean quantum cosmology (it is essential that the path integral is given by a stationary point of the Euclideanized action) which may have nothing to do with the real world, (iii) moreover if the path integral have a phase, that might eliminate the peak in the probability distribution at zero cosmological constant, i.e. at $\Lambda = 0$. In short there are too many controversies with Coleman's very speculative proposal.

One promising possibility which has not been considered so far in understanding the cosmological constant problem is the use of torsion in a framework such as the Einstein-Cartan (E-C) theory, which is natural in considering the gravitational contributions of particles with spin which is indeed a universal property of elementary particles. In fact, at sufficiently early epochs the energy content of the universe can indeed be spin dominated and the temporal evolution of the spin-density tensor is important in describing the cosmological dynamics as we shall see in section 3 [15, 16,17].

3. TORSION AND WORMHOLES

We shall first of all indicate how the antisymmetric field strength for torsion can give rise to instanton wormhole solutions. In the framework of an SL(2,C) gauge theory, the Einstein-Cartan action can be expressed as [17,18,19,20,21] (the treatment here follows closely that of ref.[17]:

$$L = -R/(16\pi G) + S_{\alpha\beta\gamma}S^{\alpha\beta\gamma}/2\pi G \tag{9}$$

where the antisymmetric field strength $S_{\alpha\beta\gamma}$ is related to the conserved current closing on the SL(2,C) algebra as

$$J^\mu = J^\mu + (1/16\pi G)\varepsilon^{\mu\alpha\beta\gamma}S_{\alpha\beta\gamma} \tag{10}$$

and is given by

$$S_{\alpha\beta\gamma} = c_\alpha \times f_{\beta\gamma} \quad \text{and} \quad f_{\beta\gamma} \cdot \bar{g} = F_{\beta\gamma} \tag{11}$$

where $\bar{g} \equiv (g_1, g_2, g_3)$ are tangent vectors to the generators of the SL(2,C) gauge group and c_α is structure constant for the group. $F_{\beta\gamma}$ in terms of the gauge fields A_β has the usual structure $F_{\beta\gamma} = \partial_\beta A_\gamma - \partial_\gamma A_\beta + [A_\beta, A_\gamma]$. $S_{\alpha\beta\gamma}$ satisfies the identity $\partial_\rho(\varepsilon^{\rho\alpha\beta\gamma}S_{\alpha\beta\gamma}) = 0$.

In the absence of matter, but with only torsion (analogous to neutrinos with spin but no mass), and with the usual Euclidean space-time metric ansatz:

$$ds^2 = d\tau^2 + a^2(\tau)d^2\Omega_3 \tag{12}$$

($d^2\Omega_3$ being the metric of S^3), the field equations reduce to [17]:

$$(da/d\tau)^2 = \left[1 - r_{min}^4/a^4\right] \tag{13}$$

where

$$r_{min}^4 = 3G^2S^2/8c^4 \tag{14}$$

S being the spin scalar.

With S identified as the basic unit $\hbar$ of spin, the minimum radius r_{min} has the value of the Planck length and is not arbitrary as in the case of solutions involving other antisymmetric field strength or scalar fields with random coupling (for instance we have found that in the case of black hole evaporation with torsion [22] effects, we are led to a final stage of a remnant mini-black-hole with angular momentum of $\hbar$). Thus we find the Coleman result without any arbitrary constants. The time $\tau = 0$ is chosen when the radius attains its minimum value r_{min}. We have therefore obtained a solution which describes tunneling from euclidean space-time to another via a baby universe of radius r_{min}, with a tunneling amplitude (since these kind of wormholes can join spaces with different topology, they represent tiny quantum fluctuations of space):

$$\psi \sim k \exp(-S_{min}/\hbar) \qquad (15)$$

where [22]
$$S_{min} = \hbar (3\pi^2/4)(r_{min}/l_{Pl}) \qquad (16)$$

With an effective cosmological constant $\Lambda(\alpha)$:

$$\Lambda(\alpha) = \Lambda_0 - \alpha k \exp(-S_{min}), \qquad (17)$$

we can arrive at Coleman's expression for the wave function of the universe:

$$\psi = \int_{-\infty}^{\infty} \frac{d\alpha}{(\pi)^{1/2}} <\exp(V_{min}\alpha)>^\beta Z(\alpha) \qquad (18)$$

where V is the volume of the 3-space and the weight $Z(\alpha)$ is given by:

$$Z(\alpha) = \exp(-\alpha^2/2) \exp(k \exp \frac{3}{8G^2\Lambda(\alpha)}) \qquad (19)$$

thus having an extremely sharp peak at $\Lambda(A) = 0$.

We can also include other gauge fields in the above action (given by eq.(9)), for instance a Yang Mills field, thus:

$$L = -(1/16\pi G) + (1/2\pi G)S_{\alpha\beta\gamma}S^{\alpha\beta\gamma} + (1/4g^2)F^\beta_{\mu\nu}F^{\mu\nu}_\beta \qquad (20)$$

($F^\beta_{\mu\nu}$ is the SU(2) Yang Mills field strength given by:

$$F^\beta_{\mu\nu} = \partial_\mu A^\beta_\nu - \partial_\nu A^\beta_\mu + \varepsilon^\beta_{ab} A^a_\mu A^b_\nu \qquad (21)$$

g is the coupling of the field and $D^\mu F^\beta_{\mu\nu} = 0$. Correspondingly we have the solution:

$$\left(\frac{da}{d\tau}\right)^2 = \left(1 - \frac{\beta_1}{a^2} - \frac{\beta_2}{a^4}\right), \qquad (22)$$

where β_1 and β_2 are constants. For $g^2 \approx \hbar c$, $\beta_1 = r^2_{min}$, $\beta_2 = r^4_{min}$ and r_{min} is identified with the Planck length as in the previous solution given by eq.(13) and (14). This then describes a wormhole with spin $\hbar$ and the charge $Gm^2 = \hbar c$. We can easily extend the above solutions (i.e. eqs.(13) and (22) to D dimensions, with similar consequences. The extension is trivial and simply involves using antisymmetric fields of higher rank and would not be dealt with here.

Apart from the usual criticism of this approach, such as the use of the Euclidean metric (it is essential that the path integral be given by a stationary point of the Euclidean action) and neglect of phases, it may be pointed out that any 3-index gauge field H can mimic a cosmological term. In general for this, the space-time dimension should be d ($\geq$ 3) and the correspondig rank of H is then (d - 1). Thus for d = 4, H is of rank 3. For instance for an action such as [17]:

$$(1/16\pi G)\int d^4x \sqrt{-g}\ (R + 2\lambda) - \int d^4x\ (1/48)\sqrt{-g}\ A^{\mu\nu\rho\sigma}A_{\mu\nu\rho\sigma}, \tag{23}$$

(where $A_{\mu\nu\rho\sigma} = \partial_{[\mu}H_{\nu\rho\sigma]}$), substituting the solution

$$\sqrt{-g}\ A^{\mu\nu\rho\sigma} = k\varepsilon^{\mu\nu\rho\sigma} \tag{24}$$

(following from $D_\mu A^{\mu\nu\rho\sigma} = 0$) into the corresponding energy-momentum tensor $T^{\mu\nu}(H)$ for the H field, as resulting from the field equations corresponding to eq.(22), we get:

$$T^{\mu\nu}(H) = - (1/2)\ k^2 g^{\mu\nu} \tag{25}$$

This is an effective Λ-term with the negative sign. In this sense one can understand how the use of the 3-index tensor $S_{\alpha\beta\gamma}$ for the torsion case, gave rise to an effective negative cosmological constant. With torsion following naturally from the spin of the particles, it is not necessary to introduce any additional antisymmetric field of rank 3 to counteract the cosmological term. However it must be noted that substituting an ansatz into the action and varying that action does not yield the same result as substituting the ansatz into the field equations [17, 23]. This seems to be a basic difficulty with wormhole type approaches. In the next section we shall consider a specific example to demonstrate the possibility of cancelling the cosmological constant term with a torsion term in the spin-dominated phase in the early universe [15.17].

4. CAN TORSION CANCEL THE COSMOLOGICAL CONSTANT IN THE EARLY UNIVERSE ?

In the Einstein-Cartan Theory, the Lagrangian is the usual scalar curvature [18,24]:

$$L_{E-C} = (-g)^{1/2} R(\Gamma) \tag{26}$$

where Γ is the non-symmetric affine connection

$$\Gamma^{\mu}_{\alpha\beta} = \begin{Bmatrix} \mu \\ \alpha\beta \end{Bmatrix} - K_{\alpha\beta.}{}^{\mu} \tag{27}$$

where the contorsion tensor $K_{\alpha\beta.}{}^{\mu}$ is related to the torsion tensor $Q_{\alpha\beta.}{}^{\mu} \equiv \Gamma^{\mu}_{[\alpha\beta]}$ by

$$K_{\alpha\beta.}{}^{\mu} = - Q_{\alpha\beta.}{}^{\mu} - Q^{\mu}_{.\alpha\beta} + Q_{\alpha.\beta}^{\ \ \mu} \tag{28}$$

The variation with respect to $g_{\alpha\beta}$ and $K_{\alpha\beta\mu}$ gives the field equations

$$G^{\alpha\beta}\left(\{\ \}\right) = \chi\,(T^{\alpha\beta} + \tau^{\alpha\beta}) \tag{29}$$

where $T^{\alpha\beta}$ is the usual energy-momentum tensor and $\tau^{\alpha\beta}$ can be considered as representing the contribution of an effective spin-spin interaction [18,24]. This is equivalent to an effective cosmological term [25]. For unpolarized spinning fluid $\tau^{\alpha\beta}$ can be decomposed in two parts: the usual canonical energy-momentum tensor for spinning particles:

$$\langle\tau^{\alpha\beta}\rangle = (1/2)\chi\,\sigma^2 u^{\alpha} u^{\beta} + (1/4)\chi\,\sigma^2 g^{\alpha\beta} \tag{30}$$

where $\quad \sigma^2 = (1/2)\langle S_{\alpha\beta} S^{\alpha\beta}\rangle \tag{31}$

and also the quadratic part corresponding to the spin-spin and spin-torsion term that is:

$$\langle\tau^{\alpha\beta}\rangle = - \chi\,\sigma^2 u^{\alpha} u^{\beta} \tag{32}$$

$S_{\alpha\beta}$ is the spin density associated with the quantum mechanical spins of elementary particles. Even if the spins are randomly oriented, the average of the spin squared terms is not zero in general [24].

As regards $T^{\alpha\beta}$ we have:

$$\langle T^{\alpha\beta}\rangle = (\rho + p) u^{\alpha} u^{\beta} - p\,g^{\alpha\beta} \tag{33}$$

Of course we are considering only the case of unpolarized spinning fluid (so there are no antisymmetric terms), so we deal only with averaged squared terms: we don't require alignment of spins [see for instance B.Kuchowicz [24]). For alignment of spins we require a very large initial magnetic field, which owing to exponential expansion would be diluted by exp[iHt] and regenerated in hadron era [26] with strong gravity. The analogy with phase transition in ferromagnetism is developed in ref.[27].

The simplest E-C generalization of standard big-bang cosmology is obtained by considering the universe filled with unpolarized spinning fluid and solving for the modified Einstein equations:

$$G^{\alpha\beta}(\{\}) = \chi \vartheta^{\alpha\beta} \qquad (34)$$

where
$$\vartheta^{\alpha\beta} = \langle T^{\alpha\beta}\rangle + \langle \tau^{\alpha\beta}\rangle$$
$$= (\rho + p - (1/2)\chi\sigma^2)u^\alpha u^\beta - (p - (1/4)\chi\sigma^2)g^{\alpha\beta} \qquad (35)$$

where ρ, p and σ depend only on time. In the comoving frame $u^\mu \equiv (0,0,0,1)$, we get the following modified field equations of the Robertson-Walker universe, which in general for $k \neq 0$ and $\Lambda \neq 0$ is of the form [16]:

$$\dot{R}^2/R^2 = (8\pi G/3)\left[\rho - (2/3)\pi G\sigma^2/c^4\right] + \Lambda c^2/3 - kc^2/R^2 \qquad (36)$$

We immediately notice that the torsion term in eq.(36), (the second term within the square brackets), is of <u>opposite</u> sign to that of the cosmological constant term. This raises the possibility that a sufficiently large spin-torsion term in early universe might cancel a correspondingly large cosmological torsion term. We shall see that this is indeed the case. For instance consider the universe at the Planck epoch when as we noted earlier the Λ term was $\approx 10^{66} cm^{-2}$, implying $\Lambda_{Pl} c^2 \approx 10^{87}$ in eq.(36). At $t_{Pl} \approx 10^{-43}$s the universe had a density of $c^5/G^2\hbar \approx 10^{93}$g cm^{-3}, and as the particle masses were $\approx 10^{19}$Gev $\approx 10^{-5}$g, the particle number density was $n_{Pl} \approx 10^{98}$cm^{-3}, so that σ, the spin density, was $\sigma_{Pl} \approx 10^{98} \cdot 10^{-27}$ (i.e. $n_{Pl}\hbar$) $\approx 10^{71}$. This gives for the term $-(8\pi G/3)(2/3)\pi G\sigma^2_{Pl}/c^4$ (i.e. the torsion term in eq.(36)) the value of $\approx -10^{87}$, which is exactly equal and of opposite sign to that of the cosmological term which is $\approx +10^{87}$, so that the two terms would have cancelled each other in the early universe at the Planck epoch [16].

We can also see that they would continue to cancel at

later epochs in the early universe. The Λ-term would evolve with temperature T as $\Lambda \propto T^2$ [28]. σ being the spin density (i.e. proportional to number density of spins) would scale as T^3 ($n \propto T^3$ for relativistic particles). So σ^2 would scale as T^6. Now in earlier work [16] it was argued that the spin-torsion coupling in the early universe was energy dependent and scaled as $G \propto T^{-2}$. So $G^2 \propto T^{-4}$, so that $G^2\sigma^2$ term would scale as T^2, which has the same dependence on T as the Λ-term so that if they cancel each other at the Planck epoch, they would also cancel at later epochs in the early universe.

Thus, in short, we have a more natural mechanism for the understanding of a vanishing Λ-term by the simple incorporation of spin-effects (a universal property of particles!) in general relativity. as regards the possible behaviour of the cosmological constnat after the hadron era, see the last section.

5. OTHER CONSEQUENCES

In another recent work [29] we have also argued that torsion can also give rise to inflation by producing an effective negative pressure term. A large cosmological term can of course also be induced in the early universe by various means, like breking of scale invariance at $E_{Pl} \approx 10^{19}$Gev and by spontaneous formation of a defect [20] in space-time which is treated as a deformable medium with elastic properties. In the geometrical description of crystal dislocations or defects it is known, for instance, that torsion plays the role of defect density [27] in the limit of dislocations having a continuous distribution. If we consider a small closed circuit and write $l^\alpha = \oint Q^\alpha_{\beta\gamma} \delta A^{\beta\gamma}$, where $dA^{\beta\gamma} = dx^\beta \wedge dx^\gamma$ is the area element enclosed by the loop and $Q^\alpha_{\beta\gamma} = \Gamma^\alpha_{[\beta\gamma]}$ is as usual the torsion associated with the connection $\Gamma^\alpha_{\beta\gamma}$. Then l^α represents the closure failure, i.e. torsion has an intrinsic geometric meaning, it represents the failure of the loop to close analogous to the crystal case, l^α having the dimensions of length.

In the above equation torsion can be related to spin, i.e. to fundamental unit of intrinsic spin ℏ, by postulating that the defect in space time topology should occur in multiples of the Planck length $L_{Pl} = (\hbar G/c^3)^{1/2}$. This is also justified by the fact that we also find wormhole solutions with torsion with a size corresponding to L_{Pl}.

It would be interesting to think of an analogue of the Cartan-Bonnet theorem for curvature, in the case of torsion: we have the result $\iint QdA$ = integer (in unit of Planck length). We would expect for a closed curve, the result $\int Qdl \geq 2\pi$, since torsion imply closure failure. These interesting topological consequences are worth developing.

The curvature associated with the defect is enormous i.e. $\approx (c/G\hbar) \approx 10$ cm same as Λ_{Pl}. A topological entity with such a large intrinsic constant curvature would expand exponentially in a de Sitter fashion with R scaling as $\exp(\sqrt{\Lambda_{Pl}} t)$. As the pressure is negative [29,30], expansion would give rise to an increasing negative energy balanced by the creation of more and more particles with positive rest mass and kinetic energy. So we have the following scenario for the formation of our universe: the exponential expansion of a single topological entity of size $\approx 10^{-33}$ cm and mass $\approx 10^{-5}$ g continues till it increases by a factor of 10^{20}, i.e. it reaches a size of $\approx 10^{-13}$ cm. The density continues to be Planck density but now the total mass of the particles created is that of the mass now in the present universe, i.e. $(10^{20})^3 M_{Pl} \approx 10^{60} M_{Pl} \approx 10^{55}$ g. The gravitational self energy of the universe is GM_U^2/l_0 with $l_0 \approx 10^{-13}$ cm. Equating this to the curvature energy $\Lambda c^2 M_U l_0^2$ gives the required initial curvature as:

$$\Lambda_0 = GM_U/c^2 l_0^3 \approx 10^{66} \text{cm}^{-2} \qquad (37)$$

which is the Planck curvature we started with! Thus the exponential expansion can create all the observed mass in the universe when the expanding radius reaches a size of 10^{-13} cm (i.e. expansion over 20 orders of magnitude, starting from the Planck length $\approx 10^{-33}$ cm) driven by the enormous negative pressure associated with the constant Planck curvature. Prior to this, the spin-torsion term is too small to counteract the cosmological term significantly.

At this stage the expansion halts because the spin-torsion interaction of the newly created particles (with a density n_{Pl}) contributes the second term in brackets in eq.

(36) which cancels the cosmological constant term driving the inflation. So the total energy is zero, as we have cancellation between the cosmological term and spin-torsion term and cancellation between the total particle energy (which is negative) and their gravitational self energy (which is negative). So we have started out with a spontaneously formed exponentially expanding topological defect and ended up creating the whole universe with a net expenditure of zero energy! (apart from quantum fluctuations!). So starting from an intrinsic spin and its associated defect we end up creating the mass observed.

The matter is still not in the form of baryons, but in the form of $\approx 10^{60} M_{Pl}$ mini-blackholes. The total entropy has increased from a value of k_B (the Boltzmann constant) for the initial defect to a value of $\approx 10^{80} k_B$. Evaporation of the $\sim 10^{60}$ black holes would further increase the entropy to $\approx 10^{87} k_B$, i.e. that of the present universe [31]. The baryon asymmetry in the evaporation could produce the observed baryon number. The universe would then continue to expand in the Robertson-Walker way, with the Λ-term now quenched by the spin-torsion interaction, halting the inflation. The present temperature of the microwave background in this picture can be shown to be [32]:

$$T_0 \simeq (m_p c^2/k_B)(8 G m_p^2/3\hbar c)^{1/3} = 2.73 \ {}^{\circ}K \qquad (38)$$

(m_p is the proton mass), agreeing well with the observed value.

At the Planck epoch, the temperature was $\approx 10^{32} \ {}^{\circ}K \approx m_{Pl} c^2/k_B$, at the hadron era $T_h \simeq m_p c^2/k_B \approx 10^{13} \ {}^{\circ}K$ and so on. Could one possibly have a residual cosmological constant? Since the universe could have a net angular momentum of ≈ 10 [33,34], the second term in eq.(36) would imply a negative cosmological constant corresponding to a density $\approx G\sigma^2/c^4 \approx 10^{-30}$g cm^{-3} not far from the critical density.

Until the hadron era we have the relation $G^2 \sigma^2 \propto T^2$ so that this temperature dependence ensures cancellation of the Λ-term induced at the various phase transitions. After the hadron era the second term in eq.(36) continues to evolve with time since the universe has a net spin density, and the present value of $G\sigma^2/c^4$ is, as we have seen, $\approx 10^{-30}$g cm^{-3}. So also the second term in eq. (36) has a very rapid fall with time.

As far as the cosmological constant, the present value would be expected to be comparable to the present curvature which is $\approx 10^{-56} \text{cm}^{-2}$. Since Λ scales as T^2 with temperature in the early universe, we would expect Λ_h in the hadron era, to be $\Lambda_{Pl}(T_h/T_{Pl})^2 \approx 10^{66} \cdot 10^{-40}$ (since $T_h \approx 10^{13}$ °K, $T_{Pl} \approx 10^{33}$ °K) that is $\Lambda_h \approx 10^{26} \text{cm}^{-2}$. This corresponds to vacuum energy densities of 10^{94}g cm^{-3} and 10^{17}g cm^{-3} in hadron era, and agrees well with what is obtained in the strong gravity picture for hadron era [18,28]. After hadron era the evolution of the cosmological constant with time would be expected to scale as t^{-2} [28]. So since $t_h \approx 10^{-23}$s, and since the present Hubble age is $\approx 10^{18}$s,, this would imply a present value of the cosmological constant of $10^{26}(10^{-23}/10^{18})^2 \approx 10^{-56} \text{cm}^{-2}$.

In some wormhole scenarios the universe could end up with a net negative cosmological constant [35].

Of course we did not invoke wormholes in this picture to achieve cancellation of the cosmological constant so the difficulties associated with the wormhole picture do not apply to this situation. It is of interest note that a recent observation [1] of number count of faint galaxies and their red shift distribution favours a low density universe ($\Omega < 0.1$) and a sizable cosmological constant.

To be consistent with inflation and dark matter ($\Omega \approx 1$), this would require the residual cosmological constant to be negative (since this can reduce Ω, i.e. $\lambda + \Omega \approx \Omega_{obs} < 0.1$. An extra cosmolgical repulsion term could affect the dynamics of large scale structures increasing their velocity dispersion so that the kinetic energy balances the extra potential term. It would be interesting to examine the consequences of this for dark matter problem.

We could also possibly have a connection between the Λ-term and the maximal entropy of a system with fixed total energy. This occurs because $\Lambda^{-1/2}$ provides a sort of infrared or long wavelength cut off to frequencies. Now for a system of total energy E the maximal entropy is given:

$$S_{max} = k_B E/\varepsilon_{min} \tag{39}$$

where ε_{min} is the minimum energy of an individual particle:

$$\varepsilon_{min} \approx \hbar c/\lambda_{max} \approx \hbar c \Lambda^{1/2} \qquad (40)$$

Now for the Planck length size defect,

$$\Lambda \approx (c^3/\hbar G)^{1/2}$$

so that
$$E \approx (\hbar c^5/G)^{1/2}$$

$$S_{max} = k_B E/\hbar c \Lambda^{1/2} = \left[k_B (\hbar c^5/G)^{1/2} \Big/ \hbar c (c^3/\hbar G^{1/2})^{1/2} \right] = k_B \qquad (41)$$

This is just the Boltzmann constant, the basic unit of entropy (see[31]). Since the present value of the Λ-term is $\Lambda_0 \approx 10^{-57}$, i.e. $\Lambda_0/\Lambda_{Pl} \approx 10^{-120}$, this would give a factor of 10^{60} increase in entropy as $S_{max} \propto \Lambda^{-1/2}$. But since the energy of the universe is $\approx 10^{60} E_{Pl}$, this would mean that the maximum possible entropy of the universe is $\approx 10^{120} k_B$. This could be of thermodynamic significance for the cosmological constant; the entropy scaling as $\Lambda^{-1/2}$, so as the curvature of the universe decreases, owing to the expansion, entropy increases.

As far as the present value of the cosmological constant is concerned, we are rather in the same situation as Eddington who remarked "I am a detective in search of a criminal - the cosmical constant. I know he exists but I do not know his appearence, for instance I do not know if he is a little man or a tall man"!

REFERENCES

1. M.Fukujta, F.Takahara, K.Yamashita and Y.Yoshii: Ap.J.Lett. 361, L1 (1990)
2. H.R.Butcher: Nature 328, 127 (1987)
3. Borgeest and S.Refsdal: Astron.Astrophys.141, 318 (1985)
4. M.Fukujta and C.Hogan: Nature 347, 120 (1990)
5. C.Sivaram: Astrophys.Space Sci.125, 189 (1986)
6. C.Sivaram: Int.J.Theor.Phys.25, 825 (1986); 26, 1625 (1987)
7. G.H.Jacoby, R.Ciardullo and H.Ford: Astrophys.J.356, 332 (1990)
8. S.Weinberg: Rev.Mod.Phys.61, 1 (1989)
9. V.de Sabbata: Giornale di Astronomia, 275 (1984); Vatican Observatory Publ., 43 (1987)
10. C.Sivaram: Astrophys.Space Sci.127, 33 (1986)

11. A.D.Dolgov: in 'The very early universe' ed.by G.Gibbons et al.Cambridge University Press 1982 pg.449
12. F.Wilczek: in 'Proc.of 21th Course of SubnuclearPhysics' ed.by A.Zichichi, Plenum Publ.Corp.N.Y.1985 pg.208
13. S.Coleman: Nucl.Phys.B307, 867 (1988)
14. A.Einstein and N.Rosen: Phys.Rev.48, 73 (1935)
15. V.de Sabbata and C.Sivaram: Astrophys.Space Sci.165, 51 (1990)
16. V.de Sabbata and C.Sivaram: Astrophys.pace Sci.158, 347 (1989)
17. V.de Sabbata and C.Sivaram: Int.J.Theor.Phys. in press
18. V.de Sabbata and G.Gasperini: 'Introduction to Gravitation' World Sci.Singapore 1985
19. C.Sivaram and K.Sinha: Nuovo Cimento Lett.13, 1357 (1975)
20. C.Sivaram and K.Sinha: Phys.Rep.51, 111 (1979)
21. M.Carmeli: Group Theory and General Relativity McGraw Hill N.Y.1984
22. V.de Sabbata, C.Sivaram, D.Wang: Ann der Phys.47,508(1990)
23. M.Duff: Phys.Lett.226B, 36 (1989)
24. F.W.Hehl, P.von der Heyde, G.D.Kerlick and J.M.Nester: Rev.Mod.Phys. 48, 393 (1976); see also B.Kuchowicz: Acta Cosmological Z.3 109 (1975)
25. C.Sivaram, K.Sinha and E.A.Lord: Nature 249, 64 (1974)
26. V.de Sabbata and M.Gasperini: Lett.Nuovo Cimento 27,133 (1980); V.de Sabbata and C.Sivaram: Il Nuovo Cimento 102B, 107 (1988)
27. V.de Sabbata and C.Sivaram: 'Torsion and Quantum effects' in volume devoted to 85th jubilee of D.D.Ivanenko, World Sci.Singapore 1991
28. C.Sivaram, K.Sinha and E.C.G.Sudarshan: Found.Phys. 6, 717 (1976)
29. V.de Sabbata and C.Sivaram: Astrophys.Space Sci.1990 in press
30. V.de Sabbata and C.Sivaram: 'Negative mass and Torsion' in volume devoted to 80th Ya.P.Terletsky jubilee, Moscow 1991
31. V.de Sabbata and C.Sivaram: 'The final state of an evaporating black hole' in "Quantum Mechanics in curved space-time ed.by J.Audretsch and Venzo de Sabbata, Plenum Publ.Corp.N.Y.1990
32. V.de Sabbata and C.Sivaram: in preparation

33. V.de Sabbata and G.Gasperini: Nuovo Cimento Lett. $\underline{25}$, 489 (1979)
34. V.de Sabbata and G.Gasperini: J.T.S.C., Journal of Tensor Society of India, Lucknow $\underline{1}$, 39 (1983); Nuovo Cimento Lett. $\underline{37}$, 605 (1983)
35. V.Rubakov: private communication

VARIATIONS OF CONSTANTS AND EXACT SOLUTIONS IN MULTIDIMENSIONAL GRAVITY

S.B.Fadeev, V.D.Ivashchuk and V.N.Melnikov

USSR State Commitee for Standards
9 Leninski Prospect, 117049, Moscow, USSR

INTRODUCTION

Constants in any physical theory characterize the stability properties of matter. With the progress of science some theories substitute others, so new constants appear, some relations between them are being established. So, the number of fundamental constants is changing. At present such choice seems more preferable: c, h, e, m_e, the Weinberg angle Θ_w, Cabbibo angle Θ_c, QCD cut-off parameter Λ_{QCD}, G, Hubble constant H, mean density of the Universe ρ, cosmological constant Λ. They may be classified as universal, as constants of interactions, and as constants of elementary constituents of matter [1].

The knowledge of constants has not only a fundamental meaning but also the metrological one. Modern system of standards is based mainly on stable physical phenomena. So the stability of constants plays a crucial role. As all physical laws were established and checked during last 2-3 centuries in experiments on the Earth and its surroundings, i.e. at a rather short space and time intervals in comparison with the radius and age of the Universe the possibility of slow variations of constants can not be excluded a priori.

Here we dwell upon mainly on the problem of the gravitational constant G stability. It is part of a very much developing field, called gravitational-relativistic metrology. It appeared due to the growth of a measuring technique precision, spread of measurements over large scales and tendency to the unification of fundamental physical interaction (see V.N.Melnikov in [1]).

Absolute value measurements of G

There are several laboratory determinations of G with precisions of 10^{-3} and only 4 at the level of 10^{-4}. They are (in $m^3 kg^{-1} s^{-2}$):

1. Facy, Pontikis, 1972 - 6,6714±0,0006
2. Sagitov et al., 1979 - 6,6745±0,0008
3. Luther, Towler, 1982 - 6,6726±0,0005
4. Karagioz, 1988 - 6,6731±0,0004

From this table it is seen that the first three contradict each other (they do not overlap within their accuracies). So the fourth experiment is in accordance with the third.

The official CODATA value of 1986
$$G = (6,67259±0,00085) \cdot 10^{-11} m^3 kg^{-1} s^{-2}$$
is based on the Luther and Towler determination. One should make a conclusion that the problem is still open and we need further experiments on the absolute value of G. Many groups are preparing them using different types of technique, among them are Karagioz (USSR) and Luo Jun (China).

There exist also some satellite determinations of G (namely GM) at the level of 10^{-8} and several geophysical determinations in mines. The last give much higher G values than the laboratory ones.

The precise knowledge of G is necessary for the evaluation of mass of the Earth, planets, their mean density and in the end for the construction of Earth models; transition from mechanical to electromagnetic units and back; evaluation of other constants through relations between them given by unified theories; finding new possible types of interactions and geophysical effects by comparing laboratory and mine determinations.

Possible variations of G

A lot of experiments were made to test possible dependence of G on electrical, magnetic phenomena, chemical and physical states of matter, temperature, radioactivity, anisotropy of space etc. They didn't give positive results concerning these variations within their accuracies (see G.Gillies in [1]) with the exception of possible variations of G with time, composition or range.

As to range or composition dependence of G (or alternatively a possible existence of new interactions - the fifth or sixth forces) the experiments are very controversial.

Among them are mine, Eotvos, Cavendish and Galileo type experiments. The search is made in the range of 1-100m, the so-called geophysical window, because in other ranges there are strong limits against possible violation of the Newton's law. Many groups in the whole world are conducting these experiments initiated by E.Fischbach, including groups from the USSR (Karagioz, Mitrofanov), China (Luo Jun), Italy (Largo experiment in Bologna) and India (Tata Institute, Bombay).

Experimental data on temporal variation of G are also contradictory. From one side there are positive van Flandern's data on the mean Moon motion at the level $\dot{G}/G \simeq 5 \cdot 10^{-11} \text{year}^{-1}$ (1976,1978). From another side we have Helling's data based on the Viking mission at the level $|\dot{G}/G| < 5 \cdot 10^{-12} \text{year}^{-1}$ and also Reasenberg's estimation at the level $|\dot{G}/G| < 5 \cdot 10^{-11} \text{year}^{-1}$ based on the same Viking data.

Great hopes were connected with Phobos-1988 mission. Due to its failure we may now wait for the next Mars mission of 1994 in order to obtain about one order improvement.

In order to explain data on possible time variations properly one should have corresponding theoretical schemes. Among such approaches we may mention different scalar-tensor theories and of course multidimensional theories. In the last case possible variations of fundamental physical constants are tests of extra dimensions.

We discuss exact cosmological solutions in

multidimensional case leading to temporal variations of G and also exact solutions in spherically-symmetrical case leading to deviations from Newton's and Coulomb's laws.

EXACT SOLUTIONS IN MULTIDIMENSIONAL COSMOLOGY

The progress in superstring theories stimulated an activity in multidimensional theories of gravitation. The idea of the time variation of the Newton's gravitational constant G originally proposed by Dirac [2] assumed a great importance with the appearance of these theories. Predictions of the multidimensional models about the time variation of G must obey the present observational upper bound [3]

$$| \dot{G}/G | \leq 10^{-11} \div 10^{-12},$$

which is a gravitational test for these models.

In [4] the "Fridman-Calabi-Yau" cosmology was considered. This cosmology model (see also [5]) is based on the ten-dimensional SO(32) or $E_8 \times E_8$ Yang-Mills-Einstein supergravity theory of a superstring origin [6,7]. It was proved that under the physical restriction on the energy density ρ and the pressure p_3 in a 3-space

$$\rho > 0 , \ p_3 \geq 0 ,$$

the solution of equation of motion with the constant radius of an internal space ($a_6(t)$=const) do not exist (the internal space pressure p_6 may be arbitrary). So, the time variation of G is an unavoidable one.

In [8] in the framework of (4+N)-dimensional cosmology with an isotropic 3-space and a Ricci-flat internal space a relation between cosmological parameters and the time variation of G was obtained. (This relation generalized that of ref. [5], obtained for superstring theories.)

In particular, for H=50km/s·Mpc it gives $|g=\dot{G}/GH|<0.2$ and for the density parameter $0.8<\Omega_0<1.2$ which agree with existence experimental data.

Now we discuss the problem of integrability of multidimensional cosmology equations.

Let us consider the metric

$$g = -e^{2\sum_{j=1}^{n} N_j x^j(t)} dt \otimes dt + \sum_{i=1}^{n} e^{2x^i(t)} g_i, \qquad (1)$$

on the manifold

$$M = \mathbb{R} \times M_1 \times \ldots \times M_n, \qquad (2)$$

where (M_i, g_i) are spaces of constant curvature, dim $M_i = N_i$, $i = 1, \ldots, n$. The Einstein equations are equivalent to the following system of equations:

$$\ddot{x}^i = \theta_i \exp(\sum_{j=1}^{n} A_{ij} x^j), \qquad (3)$$

$$E = 0, \qquad (4)$$

where $\theta_i = R[g_i]$ is scalar curvature of g_i,

$$A_{ij} = -2\delta_{ij} + 2N_j; \qquad i,j = 1, \ldots n;$$

E is energy function for the Lagrangian

$$L = 1/2 \sum_{i,j=1}^{n} G_{ij} \dot{x}^i \dot{x}^j + \sum_{i=1}^{n} \theta_i \exp(\sum_{j=1}^{n} A_{ij} x^j) \qquad (5)$$

where $G_{ij} = N_i \delta_{ij} - N_i N_j$ is minisuperspace metric. Eqs.(3)-(4) can be easily integrated at least in the following two cases
 a) all $\theta_i = 0$; b) $\theta_1 \neq 0$; $\theta_i = 0$ for $i > 1$.
In the case a) the equations of motion are integrated even for the "perfect-fluid" case with the energy-momentum tensor

$$(T^N{}_M) = \text{diad}(-\rho(t), p_1(t)\delta_1^{k_1}, \ldots, p_n(t)\delta_n^{k_n})$$

and the pressures P_i are proportional to the density $\rho > 0$
$$P_i = (1-h_i)\rho,$$
h_i are constants, $i=1,\ldots,n$ [9].
  Let $_i$ 0, $i=1,\ldots,n$. (The general case: $_i$ 0, $I = 1,\ldots,n$, $_i=0$ for $i>n$, may be reduced to this case). When

$$A_{ij} = K_{ij} v_j, \qquad (6)$$

where $v_j \neq 0$ (j=1,...,n) and (K_{ij}) is Cartan matrix for some semisimple Lie algebra, the equations of motion (2) coincide with those of the Toda lattice [10]. Listing all Cartan matrices [11], it is not difficult to check that for integers $N_i > 1$ (for $N_i = 1$ $\theta_i = 0$) the representation (6) does not take place. When N_i are arbitrary and $\sum_{i=1}^{n} N_i \neq 1$, then there are following solutions of (6):

$$\{N_1, N_2\} = \{1/3, 1/3\}, \{1/3, 1/2\}, \{1/3, 3/5\} \qquad (7)$$

corresponding to Lie algebras SU(3), SO(5) and G_2 respectively. The same result (7) may be obtained if we use the Adler-Moerbeke criteria of integrability of Toda-like systems [12]; in this case we find

$$N_j = K_j / (K_j + 2) < 1 \qquad (8)$$

$K_j \in \mathbb{N}$, j=1,...,n. This consideration may be treated as indication of possible non integrability of the above mentioned cosmological model for nontrivial θ_i ($\theta_i \neq 0$).

It should be noted that for n=2 case the cosmological model may be reduced to the Toda molecule [12]. But it is impossible to generalize this reduction prescription for n component case with n>2 [13].

The model (1) in the quantum case was considered in [14], where the Wheeler-De Witt equation was proposed and some integrable cases were considered leading to D'Alembert and Klein-Gordon equations.

EXACT SOLUTIONS IN SPHERICALLY-SYMMETRIC CASE

In [15] the Schwarzschild solution was generalized for the case of n internal Ricci-flat spaces.

This solution has the following form

$$g=-(1-L/R)^a dt\otimes dt+(1-L/R)^{-a-b}dR\otimes dR+$$

$$+(1-L/R)^{1-a-b}R^2 d\Omega^2+\sum_{i=1}^{n}(1-L/R)^{a_i}g_i, \qquad (9)$$

where $L\geq 0$; $a, a_1, \ldots a_n$ are constants obeying the restriction

$$a^2+\sum_{i=1}^{n}a_i^2 N_i+(a+\sum_{i=1}^{n}a_i N_i)^2=2 \qquad (10)$$

$b=\sum_{i=1}^{n}a_i N_i$, g_i is a metric on the Ricci-flat manifold M_i, $\dim M_i = N_i$, $i=1,\ldots,n$, and $d\Omega^2$ is the standard S^2 metric. It was shown in [15] that a horizon in the 4-dimensional section of the metric (9) at $R=L$ ($L>0$) takes place if and only if

$$a-1=a_1=\ldots=a_n=0.$$

It means that in general case we have here deviations from the Newton's law due to additional internal dimensions. The corresponding estimates may be found in [15].

This solution was also generalized for $2+d+\sum_{i=1}^{n}N_i$ dimensional case [16]. The metric of this static, spherically symmetric solution ($O(d+1)$-symmetric) is following

$$g=-(1\mp(L/R)^{d-1})^a dt\otimes dt+(1\mp(L/R)^{d-1})^{\frac{a+b+d-2}{1-d}} dR\otimes dR+$$

$$+(1\mp(L/R)^{d-1})^{\frac{a+b-1}{1-d}} R^2 d\Omega_d^2+\sum_{i=1}^{n}(1\mp(L/R)^{d-1})^{a_i}g_i, \qquad (12)$$

where

$$(d-1)(a^2+\sum_{i=1}^{n}a_i^2 N_i)+(a+\sum_{i=1}^{n}a_i N_i)^2=d \qquad (13)$$

and $d\Omega_d^2$ is the standard S_d metric.

A horizon at $R=L$ ($L>0$) in the $(2+d)$- dimensional section of the metric takes place only in the case (11), i.e. all internal space scale factors are constant [16]. (In this case we have Tangherlini solution in $(2+d)$ -section.) Here in general we also came to the violation of the Newton's law.

At present there is some interest in considering the physical models with p-adic numbers [17] instead of real ones. This interest was stimulated mainly by the pioneering works on p-adic strings [18,19]. Recently a p-adic generalization of the classical and quantum gravitational theory was defined [20] and some solutions of the Einstein equations were considered [20, 21]. In this section we consider the p-adic analog of the solution (12).

Let us briefly recall the definition of the p-adic numbers [17]. Let p be a prime number. Any rational number $a \neq 0$ can be represented in the form $a=p^k(m/n)$, where integer number m and n are not divisible by p. This norm is non-Archimedian: $|a+b|_p < \max(|a|_p, |b|_p)$. The completion of R with this norm is the p-adic number field Q_p. Any non-zero p-adic number $a \in Q_p$ can be uniquely represented as the series

$$a = p^k(a_0 + a_1 p + a_2 p^2 + \ldots)$$

where $a_0 = 1, \ldots, p-1$ and $a_i = 0, \ldots, p-1$ for $i \geq 1$.

The definitions of derivatives, manifold and tensor analysis in the p-adic case are similar to those of the real case. The power p-adic function is defined as follows:

$$(1-x)^\alpha_p \equiv \exp_p(\alpha(\log_p(1+x))), \qquad (14)$$

where $|x|_p < 1$ and $|\alpha|_p < \delta_p$. Here $\delta_p = 1$ for $p \neq 1$ and $\delta_2 = 1/2$. The definition is correct, for the functions $\exp_p$ and $\log_p$ are well defined on the discs $|x|_p < \delta_p$ and $|x_p| < 1$ respectively.

Let us consider the p-adic manifold

$$Q_p \times Q_p \times S^d \times M_1 \times \ldots \times M_n, \qquad (15)$$

where $(S^d, g_{(0)})$ is a space of constant curvature

$$R^{(0)}_{ijkl} = g^{(0)}_{ik} g^{(0)}_{jl} - g^{(0)}_{il} g^{(0)}_{jk} \qquad (16)$$

and $(M_i, g_{(i)})$ are Ricci-flat manifolds. For $R \neq 0$ and

$$|L/R|^{d-1}_p < \min(1, 1/|a_i|_p), \quad i = -2, -1, \ldots, n \qquad (17)$$

with

$$a_{-2} = \frac{a+b+d-2}{1-d}, \quad a_{-1} = a, \quad a_0 = \frac{a+b-1}{1-d},$$

the metric (12) on the manifold (15) is well defined. Then the Einstein equations for the metric (12),(17) on the manifold (15) are satisfied identically, when the parameters $a, a_1, \ldots, a_n \in Q_p$ obey the restriction (13) [16].

This can be easily checked using the identity

$$((1+x)^\alpha)' = \alpha(1+x)^\alpha / (1+x),$$

$|x|_p < 1$, $|\alpha|_p < \delta_p$ (the verification of (1) in the p-adic case is just the same as in the real one).

In d=2 case this solution was considered earlier in [21]. It was pointed that there is an infinite number of rational solutions of (13) in this case. For example, we may consider the set [21]

$$a_1 = \frac{4k}{N_1(N_1+2)k^2+1}, \quad a_i = 0, \; i > 1,$$

$$a = \frac{-2N_1 k \pm k^2 N_1 (N_1 + 2) - 1}{k^2 N_1 (N_1 + 2) + 1}, \quad k \in \mathbb{Z}.$$

In the p-adic case there exist pseudo-constant functions $C = C(R)$ such that $C'(R) = 0$ but $C(R)$ is not identically constant [17], Such functions may be used in generalization of well-known solutions of differential equations. In our case there is also a possibility for the constants $c, c_1, \ldots, c_n$ and $a, a_1, \ldots, a_n$ to be replaced by the pseudo-constants (of course, the restriction (17) should be preserved).

There is another possibility to generalize the solution (12). We may suppose that the components of the metric g_{MN} belong to some extension of Q_p. It may be quadratic extension of Q_p or even Ω_p, which is the completion of the algebraic closure of Q_p [17]. In this case the constants in (12) may belong to the extension of Q_p.

Another possible generalization of the solution (9) is the generalization on the electro-vacuum case [22]. In this case the static, spherically symmetric solutions of the field equations

$$\nabla_M F^{MN} = 0, \tag{18}$$

$$R_{MN} - (1/2)g_{MN}R = \kappa^2(F_{MP}F_N{}^P - (1/4)g_{MN}F_{PQ}F^{PQ}) \tag{19}$$

where $F_{MN} = \partial_M A_N - \partial_N A_M$ is the electromagnetic field strength, are the following

$$g = -f_1(u)dt \otimes dt + (f_1(u))^{1/(3-D)} f_2^2(u) e^{-2\sum_{i=1}^n N_i(A_i u + D_i)} du \otimes du +$$

$$+ (f_1(u))^{1/(3-D)} f_2(u) e^{-2\sum_{i=1}^n N_i(A_i u + D_i)} d\Omega^2 +$$

$$+ \sum_{i=1}^n f_1^{1/(3-D)}(u) e^{2(A_i u + D_i)} g_i, \tag{20}$$

and

$$A_0(u) = -\frac{1}{\kappa^2 B}\sqrt{C_1\frac{D-2}{D-3}}\, \mathrm{cth}[\sqrt{C_1\frac{D-3}{D-2}}\,(u-u_1)] + \varphi_1 - \tag{21}$$

$$A_i = 0,\ i \geq 1,$$

where $D = 4 + \sum_{i=0}^n N_i$,

$$f_1(u) \equiv C_1/\kappa^2 B^2 \mathrm{sh}^2(\sqrt{C_1\frac{D-3}{D-2}}\,(u-u_1)),$$

$$f_2(u) \equiv C_2/\mathrm{sh}^2(\sqrt{C_2}\,(u-u_2)),$$

and $D_i, \varphi_0, u_1, u_2, B \neq 0$ are arbitrary constants and the constants C_1, C_2, A_i obey the following relation

$$2C_2 = C_1 + (\sum_{i=1}^n N_i A_i)^2 + \sum_{i=1}^n N_i A_i^2.$$

For $N \to 0$, $D \to 4$ we have the well-known Reissner-Nordstrom solution.

Now we discuss another possibility to generalize the gravitational theory. Recently great interest has been raised in quantum algebras(see, e.g. review [23]). The quantum algebras appeared in the models of quantum field theory, string theory, etc. [24]. Noncommutative geometry leads to quantum algebras as well [25,26].

In a number of models the transition from $U(g)$ to $U_q(g)$, where g is Lie algebra, $U_q(g)$ is a deformed universal enveloping of g ($U_{q=1}(g)=U(g)$), q is a deformation parameter, results effectively in a q-deformation of derivatives and integral [27]. Thus, for example, the derivative of the function f at the point t is modified as follows

$$(\partial f)(t) \to (D^{(q)}f)(t) = \frac{f(qt)-f(q^{-1}t)}{(q-q^{-1})t} \quad (23)$$

where $q \in \mathbb{R}$, $q^2 \neq 1$. Nonlocal "derivative" (23) for $q \to 1$ coincides with the ordinary one $(\partial f)(t)$. In noncommutative geometry [26] the parameter q is a noncommutativity parameter:

$$\hat{x}_i \cdot \hat{x}_j = q \hat{x}_j \cdot \hat{x}_i, \quad i<j .$$

q-deformed strings have been studied previously in [27,28].

Here a deformation of the simplest cosmological (Kasner) model is considered. For the metric

$$g = -b_1(t) \ldots b_n(t) dt \otimes dt + \sum_{i=1}^{n} b_i(t) dx_i \otimes dx_i \quad (24)$$

Einstein equations are of the form

$$\dot{h}_i = 0,$$

$$\left(\sum_{i=1}^{n} h_i\right)^2 - \sum_{i=1}^{n} h_i^2 = 0, \quad (25)$$

$$h_i = b_i^{-1} \dot{b}_i, \quad i=1,\ldots,n.$$

Solution (25) is trivial

$$h_i = C_i, \quad \left(\sum_{i=1}^{n} C_i\right)^2 - \sum_{i=1}^{n} C_i^2 = 0, \quad (26)$$

$$b_i = d_i \exp(C_i t),$$

where $d_i \neq 0$, C_i are constants, $i=1,\ldots,n$.

In the q-deformed case equations (25) are modified

$$D^{(q)}h_i = 0, \quad (\sum_{i=1}^{n} h_i)^2 - \sum_{i=1}^{n} h_i^2 = 0, \qquad (27)$$

$$h_i = b_i^{-1} D^{(q)} b_i, \quad i=1,\ldots,n.$$

Solving (27), we obtain

$$h_i = C_i, \quad (\sum_{i=1}^{n} C_i)^2 - \sum_{i=1}^{n} C_i^2 = 0, \qquad (28)$$

$$b_i = d_i \exp_q(C_i t),$$

$d_i \neq 0$, $i=1,\ldots,n$. In (28) deformed exponent is present

$$\exp_q z = \sum_{n=0}^{\infty} z^n / n_q! \qquad (29)$$

Here $n_q! = n_q \cdot \ldots \cdot 1_q$ for $n \geq 1$, $0_q! = 1$ and $n_q = (q^n - q^{-n})/(q - q^{-1})$. The convergency radius of the series in (29) is equal to $+\infty$. Function (29) may be expressed in terms of the q-deformed hypergeometrical function ${}_1\Phi_2$ [29].

REFERENCES

[1] Gravitational Measurements, Fundamental Metrology and Constants. Eds.V.de Sabbata and V.N.Melnikov, Kluwer Acad.Publ.Dordtrecht,Holland,1988.
[2] P.A.M.Dirac, Nature(London),139,321(1937).
[3] R.W.Hellings et.al., Phys.Rev.Lett.,51,1609(1983).
[4] V.D.Ivashchuk and V.N.Melnikov, Nuovo Cimento B, 102, 131(1988).
[5] Y.-S.Wu and Z.Wang, Phys.Rev.Lett.,57,1978(1986).
[6] G.F.Chapline and N.S.Manton, Phys.Lett.B,120,105(1983).
[7] P.Candelas,G.T.Horowitz,A.Strominger and E.Witten, Nucl. Phys.B,256,46(1985).
[8] K.A.Bronnikov, V.D.Ivashchuk and V.N.Melnikov, Nuovo Cimento B,102,209(1988).
[9] V.D.Ivashchuk and V.N.Melnikov, Phys.Lett.A, 136, 465(1989).
[10] M. Toda, Theory of Nonlinear Lattices (Berlin, Springer, 1981); B.Kostant, Adv.Math. 34(1979)195; M.A.Olshanetsky and A.M.Perelomov, Phys.Rep.71(1981)313.

[11] N.Bourbaki, Groupes et Algebres de Lie, Ch.4-6, Hermann, Paris, 1968.
[12] G.W.Gibbons and K.Maeda, Nucl.Phys.B, 298, 741(1988).
[13] V.D.Ivashchuk, V.N.Melnikov, Chinese Phys. Lett., 7, 97(1990).
[14] V.D.Ivashchuk, V.N.Melnikov and A.I.Zhuk, Nuovo Cimento B, 104, 575(1989).
[15] K.A.Bronnikov, V.D.Ivashchuk and V.N.Melnikov, Problems of Gravitation (Plenary Reports of YII Soviet Grav.Conf.), Erevan, ErGU, 1989, p.70; Nuovo Cimento B, 1990, to appear.
[16] S.B.Fadeev, V.D.Ivashchuk and V.N.Melnikov, Nuovo Cimento B, to be published.
[17] N.Koblitz, P-adic numbers, p-adic analysis and zeta-functions, Springer, New York, 1977.
[18] I.V.Volovich, Class.Quantum Grav., 4 (1987) L83.
[19] P.G.O.Freud and M.Olson, Phys. Lett.B, 199 (1987) 186.
[20] I.Ya.Aref'eva, B.Dragovich, P.Frampton and I.V.Volovich, Wave Function of the Universe and p-adic Gravity, Steklov Mathematical Institute Preprint, 1990.
[21] V.D.Ivashchuk, V.N.Melnikov and S.B.Fadeev, National Workshop on Gravitation and Gauge Theory, Yakutsk, to be published, 1990.
[22] S.B.Fadeev, V.D.Ivashchuk and V.N.Melnikov, Chin. Phys. Lett., 1991, to be published.
[23] S.Majid, Int.J.Mod.Phys.A, 5 (1990) 191.
[24] L.Alvarez-Gaume, C.Gomez and G.Sierra, Duality and Quantum Groups, CERN-TH.5369/89.
[25] A.Connes, Non-commutative Differential Geometry, IHES 62(1986).
[26] J.Wess and B.Zumino, Covariant differential calculus on the quantum hyperplane, CERN-TH-5697/90.
[27] D.Bernard and A.Leclair, Phys.Lett.B, 227 (1989) 97.
[28] H.J.De Vega and N.Sanchez, Phys.Lett.B, 216 (1989) 97.
[29] G.E.Andrew, q-series, their development and application in analysis, number theory, combinatorics, physics and computer algebra, AMS Regional Conference Series 66 (1986).

LARGER SCALE STRUCTURE IN THE LYMAN-ALPHA ABSORPTION LINES

Li-Zhi Fang

Institution for Advanced Study
Princeton, N J

1. INTRODUCTION

The motivation for studying the large scale structure of the Lyα forest is twofold. First, by comparing the large scale clustering of the Lyα forest with that of high redshift quasars, one can learn whether the biased clustering has taken place during the formation of the large scale structure in the era corresponding to the high redshift considered. Second, by comparing the Lyα forest's structure with that of galaxies, we can learn how early in the universe the large scale structure observed locally can be traced.

The redshift distribution of Lyα absorbers has been studied since 1980 (Sargent et al. 1980). Almost all results drawn from two-objects correlation studies have failed to detect any significant correlation among the absorption line redshifts on scale from ~ 300 to $30,000$ km s^{-1}. The only exception is provided by Webb (1987), who finds that the correlation function analysis gives positive evidence for weak clustering on a velocity scale of 50-290 km s^{-1}.

Nevertheless, we cannot confidently conclude that no large scale structure exists in the Lyα forest. As Liu and Jones (1990) have shown, the two-point correlation function is not the best tool for measuring line clustering. Some information on clustering, especially at very large scale, would be lost in correlation function analysis due to the effects of finite sample size and noise. Moreover, large scale structure would be erased in those Lyα forest samples, which include compilation of Lyα absorption lines of quasars located far away from each other on the sky.

Therefore, what the possible large scale structure is in the Lyα forest is still an open question. In particular, a study of samples of dense Lyα lines from individual quasars is required. Recently, using very deep pencil-beam surveys, Broadhurst et al (1990) found an excess correlation and an apparent periodicity in the galaxy distribution with a characteristic scale of 128 $h_0^{-1} Mpc$, h_0 being the Hubble constant in unit of 100 km $s^{-1} Mpc^{-1}$. This result strengthens the interest of a search for the existence of large scale structure in the Lyα forest of individual quasars. Because the Lyα forest of individual quasars is also a sample pencil-beam nature and suffering less selection effect, it provides a more straightforward comparison between the clustering of high and low redshift objects in a pencil-like region.

In this paper we start with a statistical analysis of the cumulative distribution of absorber pair intervals. We find that the most dense sample of Lyα forests given by Crotts (1989) significantly deviates from a uniform distribution on at least the scale 30-50 $h_0^{-1} Mpc$. A further power spectrum analysis shows that a periodic component with wavelength of about 80 $h_0^{-1} Mpc$ exists in the spatial distribution of Lyα absorbers, but at a lower confidence level. The significance and implications of those inhomogeneities are also discussed(Fang, 1991).

2. SAMPLE OF Lyα ABSORPTION LINES

In order to study large scale structure in a pencil-like region, Lyα forest samples drawn from a compilation of Lyα absorption lines in quasars located far away from each other on the sky are obviously not appropriate. One needs samples of Lyα forest from individual quasars or quasar groups sitting within a small region of the sky.

We choose the sample of Crotts (1989), because it is perhaps the most dense sample of Lyα forests distributed in a pencil-like region with redshifts 2-2.5. Crotts' sample comes from 1.8 Å FWHM absorption spectra of four quasars, Nos 76, 77, 78 and 79, in Sramek and Weedman's (1978) survey. The angular separations among the first three quasars are 147" (SW 76 and 77), 127" (SW 76 and 78), and 177" (SW 77 and 78). SW 79 sits approximately 6'.2 away from the center of the triangle formed by the first three.

Table 1. The Lyα forest sample

quasar	z_{em}	z_{min}	n(total)	$n(W > 0.2 Å)$
SW 76	2.467	1.9397	38	35
SW 77	2.521	1.9424	85	47
SW 78	2.607	1.9659	42	36

Since the redshift of SW 79 is too low compared with the first three and its position is also far away from the others, we will only analyze, in the following, the Lyα forests of SW 76, 77 and 78. Table 1 shows the basic data, including the name of the quasars, the emission line redshift, the lower limit of Lyα absorption redshifts, the total number of Lyα absorption lines, and the number of lines with a rest frame equivalent width $W > 0.2 Å$.

3. STATISTICAL ANALYSIS AND LARGE SCALE STRUCTURE

It has been known that the relationship between the spatial and velocity (redshift) distributions of absorbers is not straightforward. For a one-dimensional (pencil-beam) sample, the peculiar motions could strongly smear out the spatial clustering, and result in a weak clustering of objects with small velocity separation. Therefore the clustering strength of the absorbers at small velocity separation cannot be meaningfully compared with that derived from three-dimensional surveys for other classes of objects (e.g. quasars). However, on the scale larger than about 10 $h_0^{-1} Mpc^{-1}$, we may neglect the influence of peculiar motions. In these conditions, it seems safe to use straightforwardly the redshift distribution to derive the spatial clustering. Therefore, we will directly study the large scale distribution of Lyα absorbers in comoving space. When $q_0 = 1/2$, the comoving distance of an absorber at redshift z is given by

$$d = 2(c/H_0)[1 - (1+z)^{-1/2}] \tag{1}$$

The influence of the particular choice made for q_0 will be discussed later. From eq.(1) we find that the absorbers in the three pencil samples of Table 1 distribute over a straight line with lengths 270 $h_0^{-1} Mpc$, 298 $h_0^{-1} Mpc$ and 324 $h_0^{-1} Mpc$, respectively. The mean separation between adjacent absorbers is in the range of 2 $h_0^{-1} Mpc$ (sample of all lines of the three quasars) to 9 $h_0^{-1} Mpc$ (sample of $W > 0.2 Å$ lines of SW 78). Therefore, the sample defined in Sec. 2 can be used to probe the structure at scales larger than 20 $h_0^{-1} Mpc$ and less than 150 $h_0^{-1} Mpc$.

3.1 Comparison with a uniform distribution

Crotts (1989) has already performed a clustering analysis of the samples of Sec. 2 by means of two-point correlation functions. He concluded that there is no evidence for structure except for velocity splitting less than 100 $km\ s^{-1}$. Even though the two-point function may overlook some large scale clustering, Crotts' result means, at least, that the Lyα forests of the given sample does not deviate very much from a uniform distribution. Therefore, we will first test the significance level of a disproof of the null hypothesis that the Lyα forest distributes uniformly.

For detecting a weak deviation from uniform or random distributions, the method involving the nearest neighbor line interval distribution (Liu and Jones 1990; Ostriker et al. 1988; Duncan et al. 1989) is more effective. A non-random distribution of neighbor line intervals may not lead to an observable line correlation function in redshift space. The advantage of the neighbor line interval method is that it enables one to use a statistical measure involving the overall distribution. The Kolmogorov-Smirnoff (K-S) statistics computed for the maximum difference between the observed and theoretical cumulative distributions of neighbor line intervals is a measure of the overall sample, but do not give information related to specific scales. One can then expect that it may be effective in finding weak clustering overlooked by correlation function analysis.

We then first study the statistical distribution of absorber pair separation (not only the nearest neighbor one). Let us consider a pencil-like sample consisting of n objects distributed over a straight line with length D. The number of objects in the distance interval from x to $x+dx$ is given by $dN = F(x)dx$, $F(x)$ being the distribution function. The number of pairs with separation in the range of $D_L < x < d$ is then given by

$$n(<d, D_L) = \int_{D_L}^{d} \int_{0}^{D-x'} F(x)F(x+x')dx dx' \tag{2}$$

The normalized cumulative distribution of pairs with separation from D_L to D_U can be expressed as follows

$$S(<d, D_L, D_U) = N(<d, D_L)/N(<D_U, D_L) \tag{3}$$

For a uniform distribution, i.e. $F(x) = \text{const}$, we have

$$S(<d, D_L, D_U) = [(D - D_L)^2 - (D - d)^2]/[(D - D_L)^2 - (D - D_U)^2] \tag{4}$$

Now one can test the null hypothesis on scales ranging from D_L to D_U by using the K-S statistics for $L(D_L, D_U)$, which is the maximum value of the absolute difference between the cumulative distributions of pair numbers given by the observed sample and eq.(4). What makes the K-S statistics very useful in our context is that, from $L(D_L, D_U)$, we can estimate the probability $P(D_L, D_U)$ that the two distributions given by observed sample and eq.(4) are drawn from the same parent distribution. Thus a significance level of the disproof of the null hypothesis can be obtained.

The results of the K-S probabilities $P(D_L, D_U)$ for SW 76 and 78 samples show that, on scales of 20 $h_0^{-1}Mpc < D_L < 160\ h_0^{-1}Mpc$, there are no significant departures from a uniform distribution. But for SW 77 there is a deviation on the scale of 20 $h_0^{-1}Mpc < D_L < 50 h_0^{-1}Mpc$. The significance levels are 99.5% for the sample consisting of all absorption lines and 99% for the $W > 0.2\text{Å}$.

We also calculated the K-S probabilities for the difference between the observed sample and a random distribution of $F(x)$. In this case, $F(x)$ is generated from random uniform distribution of n objects on a line with length D, and n and D having the same values as the sample being tested. The parameter $L(D_L, D_U)$ is now the average of maximum absolute differences between the observed sample and 30 randomly distributed samples. The main features of $P(D_L, D_U)$ are completely the same as above. Therefore, the first conclusion of this paper is: the distribution of SW 77 Lyα forest does significantly deviate from a uniform model on a scale of 30-50 $h_0^{-1}Mpc$.

We also calculated the difference of the pair numbers of sample SW 77 from that given by uniform and random models in each distance bin. It showed that there is, indeed, a slight deficit of pairs with separation in the range 30-50 $h_0^{-1}Mpc$. One can also find a deficit of pairs at 200 $h_0^{-1}Mpc$ and an excess at 250 $h_0^{-1}Mpc$.

3.2 Influence of the dependence of density on redshift

It is well known that the Lyα line number per unit redshift interval dN/dz increases with redshift according to a power law

$$\frac{dN}{dz} \propto (1+z)^\gamma \tag{5}$$

with $\gamma \sim 2$ (Murdoch et al, 1986). Considering this point, the mean distribution of Lyα absorbers can no longer be modeled as a uniform or random distribution as we did in previous section. Nevertheless, since the length of the sample pencils is not too large, the influence of eq.(5) on previous statistics can simply be taken into account by replacing the uniform distribution function $F(x) = \text{const}$ by

$$F(x) = \text{const}(1 + ax), \quad 0 < x < D\ h_0^{-1}Mpc. \tag{6}$$

From eq.(5), one can find the order of the small parameter a in the correction term of eq.(6) given approximately by

$$a \sim D^{-1}\Delta \ln(dN/dx) \sim D^{-1}(\gamma + 1.5)\Delta z/(1+z). \tag{7}$$

where we used $dN/dx = (dN/dz)(dz/dx)$ and $dz/dx = (H_0/c)(1+z)^{3/2}$ when $q_0 = 1/2$.

Therefore, in the case of the considered samples, i.e. $z \sim 2.3$, $\Delta z \sim 0.5$ and $D \sim 300 h_0^{-1} Mpc$, we have $a \sim 10^{-3} Mpc^{-1}$. For the total line and $W > 0.2 Å$ line samples of SW 77, the precise values of parameter a are, respectively, $7.7 \cdot 10^{-4}$ and $1.8 \cdot 10^{-3}\ Mpc^{-1}$.

Substituting eq.(6) into eq.(2), the cumulative probability distribution now writes

$$S(<d, D_L, D_U) = \frac{(D-D_L)^2 - (D-d)^2}{(D-D_L)^2 - (D-D_U)^2} \cdot \frac{6(1+aD) + a^2[(D-D_L)^2 + (D-d)^2]}{6(1+aD) + a^2[(D-D_L)^2 + (D-D_U)^2]}$$

The correction term in $S(<d, D_L, D_U)$ is of the order of a^2. We can therefore expect that the statistical results of section 3.1 will not be notably influenced by the redshift dependence of the line density.

The influence of q_0 can also be described by eq.(6). The position x of an absorber in a $q_0 = 1/2$ universe will become $x'(x, q_0)$ in a q_0 universe. A uniform distribution $F(x) = $ const in a $q_0 = 1/2$ universe corresponds then in a q_0 universe to a non-uniform distribution $F(x') = $ const $\cdot dx/dx'$. Since $dx/dx' = (1+2q_0 z)^{1/2}/(1+z)^{1/2}$, $F(x')$ can be approximated by eq.(6) and parameter

$$a \sim (1+z_{min})[2q_0(1+z_{min})^{1/2}(1+2q_0 z_{min})^{-1/2} - 1](H_0/c)/2. \tag{8}$$

For $0 < q_0 < 2$, a is also of the order of $10^{-3}\ Mpc^{-1}$. Therefore, the previous result is almost totally independent of the value of q_0.

3.3 Periodicity

Why is there a deficit of pairs with separation around 40 $h_0^{-1} Mpc$? If we consider this deviation together with the less significant 200 $h_0^{-1} Mpc$ deficit and the 250 $h_0^{-1} Mpc$ excess, a possible answer is the existence of some periodicity in the clustering of Lyα absorbers. A periodic inhomogeneity with wavelengths of about 80 $h_0^{-1} Mpc$ would naturally lead to the above-mentioned excess and deficits. Therefore, it is interesting to find the confidence level of such a periodicity. The results of a power spectrum analysis (PSA) of the three samples showed a peak at wavelength 83 $h_0^{-1} Mpc$ for the sample of SW 77 $W > 0.2Å$. The confidence level is 98%. This periodicity can, in fact, already be identified in the redshift distribution of $W > 0.2Å$ Lyα lines (Figure 9 of Crotts, 1989) which shows two dips at redshifts $z = 2.22$ and $z = 2.38$. The corresponding wavelength in comoving space is 80 $h_0^{-1} Mpc$, i.e. the same as the one given by the PSA. No periodicities can be evidenced in the samples of SW 76 and 78.

3.4 Clustering of composed Lyα forest samples

A problem which arose in previous results is why the clustering characteristics are so different among the three Lyα forests of SW 76, 77 and 78. Since the proper scale of clustering in SW 77 is about 15 times larger than the proper distance between lines of sight towards SW 76, SW 77 and SW 78 at the redshifts considered, the structure of the three Lyα forests should not be very much different from each other. However, statistical results invariably shows that the distributions of Lyα lines in both SW 76 and 78 are almost uniform on a large scale.

This contradiction can further be checked by studing the clustering of composite samples, which consist of the Lyα lines of SW 77 plus 76 or SW 77 plus 78. We have the same reason as above to think that the clustering of the composed samples will also be the same as for SW 77. The result is that, even though no structure is noticeable in the SW 78 Lyα forest, the clustering of the SW 77 plus SW 78 sample is more remarkable than for the SW 77 one alone. The sample including SW 77 and 76 lines shows, however, weaker clustering. These results may show that all the inhomogeneities of SW 77 are just due to random fluctuations. It could also be explainable if the large scale structure have the shape of filaments and pancakes. Of course, a somewhat larger sample concerning groups of quasars is needed before reaching a definite conclusion. Therefore, the problem still partially remains.

4. CONCLUSION

The K-S probabilities regarding the homogeneity of the distribution of Lyα lines of the observed samples reveal that, at a $2-3\sigma$ statistical significance level, the most dense sample of SW 77, deviates from a uniform distribution due to a deficit of pair with separation in the range of 30-50 $h_0^{-1} Mpc$. A periodic component with wavelength 80 $h_0^{-1} Mpc$ has also been identified at a confidence level of 98% in the distribution of SW 77 Lyα forest. Both inhomogeneities may be related to each other, because one consequence of the 80 $h_0^{-1} Mpc$ periodicity is the deficit of line pairs with separation 30-50 $h_0^{-1} Mpc$.

These results shows that the general feature of the clustering of Lyα absorbers is not very much different from high redshift ($z > 2$) quasars. That is, the distribution of Lyα absorbers is also non-uniform but weakly clustered on scales as large as about 100 $h_0^{-1} Mpc$ (Chu and Fang, 1987; Fang 1989).

The amplitude of the quasar correlation function at 10 $h_0^{-1} Mpc$ is much higher than that of the Lyα forest (Shaver, 1988). On this scale, however, the clustering of a Lyα forest is strongly weakened by line blending. Therefore, the clustering amplitudes at about 10 $h_0^{-1} Mpc$ cannot be used as evidence for differences in the clustering between quasars (high-mass objects) and Lyα absorbers (low-mass objects). At larger scales (> 20 $h_0^{-1} Mpc$), no available data on the clustering amplitudes of quasars can be found. But we know at least, that high redshift quasars show a deviation from a random distribution with about the same significant level as for the Lyα forest (Chu and Fang, 1987). Moreover, in spite of the lack of a consistent conclusion from different studies on the periodicity in the distribution of emission line redshifts of quasars, we can at least say that the confidence level of quasar's periodicity given by power spectrum analysis is not higher than for the Lyα forest. Therefore, our result may indicate that the formation of structure in the universe had not yet undergone a biased clustering on the comoving scales of > 20 $h_0^{-1} Mpc$ until the era of $z \sim 2$. The absence of any systematic difference between the clustering strengths of the samples consisting of total lines and $W > 0.2 \text{Å}$ lines also shows no biased clustering with respect to the parameters related to the rest frame equivalent widths of absorbers.

Our results also imply that the periodicity found recently from pencil-beam surveys of $z < 0.5$ galaxies may already exist in the high redshift era, but with very low amplitude. Of course, more statistical analyses on somewhat more qualified sample of Lyα forest is needed for a definite statement. Especially, the influence of the misidentifying Lyα line with Lyβ or Lyγ lines in the sample of Crotts can only be avoid by sample with high resolution.

REFERENCES

Broadhurst, T.J., Ellis, D.C., Koo, D.C. and Szalay, A.S.: 1990, *Nature*, **343**, 726.
Chu, Y.Q. and Fang L.Z.: 1987, in *Observational Cosmology*, eds. A.Hewitt,
　G. Burbidge and L.Z. Fang, Reidel, Dordrecht.
Crotts, A.P.S.: 1989, *Astrophys. J.* **336**, 550.
Fang, L.Z.: 1989, *Inter. J. Mod. Phys.* **A4**, 3477.
Fang, L.Z.: 1991, *Astr. Astrophys.* **244**, 1.
Liu, X.D. and Jones, B.J.T.:1990, *Monthly Notices Roy. Astron. Soc.* **242**, 678.
Murdoch, H.S., Hunstead, R.W., Pettini, M. and Blades, J.C.: 1986, *Astrophys. J.* **309**, 19.
Ostriker, J.P., Bajtlik, S. and Duncan, R.C.: 1988, *Astrophys. J.* **327**, L35.
Sargent, W.L.W., Young, P., Boksenberg, A. and Tyler, D.: 1980, *Astrophys. J. Suppl.* **42**, 41.
Shaver, P.A.: 1988, in *Larger Scale Structure of the Universe*, eds.J. Audouze, M-C. Pelletan and A. Szalay, Kluwer, Dordrecht.
Sramek, R.A. and Weedman, D.W.: 1978, *Astrophys. J.* **221**, 468.
Webb, J.K.: 1987, in *Observational Cosmology*, eds. A. Hewitt, G.Burbidge and
　L.Z. Fang, Reidel, Dordrecht.

NULL SURFACE CANONICAL FORMALISM

J.N. Goldberg

Department of Physics, Syracuse University
Syracuse, NY 13244-1130

D.C. Robinson and C. Soteriou

Department of Mathematics, King's College
Strand, London WC2R 2LS

1. Introduction

More than 20 years ago, Peter Bergmann together with one of us (JNG) considered the problem of constructing the canonical formalism for general relativity on a null cone. The motivation for doing so came from the difficulty in constructing the observables which could then become the basic operators in a quantum theory of gravity. The analysis of Bondi[1], Sachs[2], and Newman-Unti[3] of the Einstein equations in the vicinity of null infinity indicated that the constraint equations were easy to integrate and that the data to be specified freely was easily recognizable. On the outgoing null cone, the geometry of the 2-surface foliation is given and at null infinity one gives the news function at all times and the mass aspect and the dipole aspect at one time. This result gave hope that in the canonical formalism one would be able to recognize the appropriate variables for the quantum theory.

A few years ago, JNG returned to this problem. He introduced the Bondi-Sachs coordinate conditions and then found the Dirac brackets for the metric of the 2-surface foliation of the outgoing null cone[4,5]. At the same time, Charles Torre[6] was working on the same problem using a tetrad based on the "2 by 2" formalism introduced by d'Inverno with Stachel[7] and with Smallwood[8]. He did not introduce any coordinate or tetrad conditions and was able to work out the constraint algebra, but he did not construct the Dirac brackets. Work on both approaches stopped at that point because the constraints appeared to be far too complicated even in this formalism.

Using the new variables of Abhay Ashtekar[9], the Hamiltonian of general relativity is a polynomial of the fourth degree. It seems reasonable to return to this problem to see whether the focus on self-duality imposed by the new variables would lead to a simplification. This work is still incomplete. But, because this problem is of interest to Peter Bergmann, for this celebration of his 75th birthday, we are presenting a sketch of our method and results. The following section is devoted to setting up the formalism and the Lagrangian although conventions and some definitions are relegated to the Appendix. Section 3 sets up the Hamiltonian and discusses the constraints. Some brief concluding remarks are made in the final section.

2. The Lagrangian

We consider the connection and the components of a tetrad to be independent configuration space variables. Therefore, we follow Samuel[10] and Jacobson and Smolin[11] in writing the Lagrangian in terms of the self-dual Riemann tensor,

$$S = i \int {}^+R_{\alpha\beta\mu\nu}\Sigma^{\alpha\beta}{}_{\rho\sigma}\eta^{\mu\nu\rho\sigma}d^4x \tag{2.1}$$

where

$$ {}^+R_{\alpha\beta\mu\nu} = \tfrac{1}{2}(R_{\alpha\beta\mu\nu} - \tfrac{1}{2}i\epsilon_{\alpha\beta\gamma\delta}R^{\gamma\delta}{}_{\mu\nu}) \tag{2.2}$$

The self-dual 2-forms are defined in terms of a null tetrad. Conventions on notation and properties of the null tetrad, the self-dual bivectors, the definition of the connection coefficients, and the definition of the Riemann tensor are given in the Appendix. Note in particular that we introduce upper case Latin indices with the range $(\mathbf{1,2,3})$ to label the bivectors. The action then takes the form

$$S = -2i \int R_{\mathbf{A}\mu\nu}\Sigma^{\mathbf{A}}{}_{\rho\sigma}\eta^{\mu\nu\rho\sigma}d^4x \tag{2.3}$$

In Eq. (A.7) of the Appendix, the self-dual Riemann tensor is expressed in terms of the self-dual components of the connection.

To construct the Hamiltonian, we do a 3+1 decomposition in the action. Thus (2.3) becomes

$$S = -4i \int \{R_{\mathbf{A}0m}\Sigma^{\mathbf{A}}{}_{\rho\sigma} + R_{\mathbf{A}\rho\sigma}\Sigma^{\mathbf{A}}{}_{0m}\}\eta^{0mrs}d^4x \tag{2.4}$$

We separate the connection into its pull-back to the null surface, $A^{\mathbf{A}}$, and the portion which leads off of the surface, $B^{\mathbf{A}}$:

$$\mathcal{G}^{\mathbf{A}} =: A^{\mathbf{A}}{}_r dx^r + B^{\mathbf{A}} du \tag{2.5}$$

We substitute (2.5) into (A.7), drop a divergence, and get from (2.4)

$$S = -2i \int \{2A^{\mathbf{A}}{}_{i,0}V_{\mathbf{A}}{}^i - N\mathcal{H} - N^i\mathcal{H}_i - B^{\mathbf{A}}G_{\mathbf{A}}\}d^4x \tag{2.6}$$

Here we have introduced

$$\begin{aligned}
V_{\mathbf{A}}{}^i &= v(V_1{}^i, 0, V_3{}^i), \quad (v = det(v^\alpha{}_i))\\
\mathcal{H} &= v_2^j(V_1{}^i R^2{}_{ij} + V_3{}^i R^1{}_{ij})\\
\mathcal{H}_i &= R^{\mathbf{A}}{}_{ij}V_{\mathbf{A}}{}^j\\
G_{\mathbf{A}} &= -2D_j V_{\mathbf{A}}{}^j
\end{aligned} \tag{2.7}$$

The covariant derivative is defined in Eq. (A.8). Note that the $V_A{}^i$ is a vector density of weight one.

3. The Canonical Formalism

From Eq (2.8) we see that the momenta conjugate to the configuration space variables

$$A^A{}_i, \; v_2{}^i, \; B^A, \; N, \; \text{and} \; N^i$$

are

$$-4iV_A{}^i, \; \pi^2{}_i, \; P_A, \; p, \; \text{and} \; p_i,$$

respectively. Note that these momenta all have density weight one with respect to diffeomorphisms on the null surface. The 38 phase space variables satisfy the standard "equal time" Poisson bracket relations on the null cone $u = $ constant:

$$\begin{aligned}
\{A^A{}_i(x), V_B{}^j(x')\} &= \tfrac{i}{4}\delta^A{}_B \delta_i{}^j \delta^3(x-x'), \\
\{V_2{}^i(x), \pi^2{}_j(x')\} &= \delta^i{}_j \delta^3(x-x'), \\
\{B^A(x), P_B(x')\} &= \delta^A{}_B \delta^3(x-x'), \\
\{N(x), p(x')\} &= \delta^3(x-x'), \\
\{N^i(x), p_j(x')\} &= \delta^i{}_j \delta^3(x-x').
\end{aligned} \quad (3.1)$$

The canonical variables also satisfy the following 13 primary constraints:

$$V_2{}^i = 0, \; \pi^2{}_i = 0, \; P_A = 0, \; p = 0, \; p_i = 0 \quad (3.2).$$

With the addition of the primary constraints, the Hamiltonian becomes

$$H = \int \{N\mathcal{H} + N^i \mathcal{H}_i + B^A G_A + \alpha_i V_2{}^i + \beta^i \pi_i^2 + \alpha^A P_A + \gamma p + \gamma^i p_i\} d^3x \quad (3.3)$$

To insure the propagation of the primary constraints, we set to zero their Poisson brackets with the Hamiltonian. This leads to the following set of secondary constraints:

$$\begin{aligned}
\chi^i &:= 2D_j(NM^{2A}{}_2 v_2{}^{[i} V_A{}^{j]}) - N^{[i} V_2{}^{j]}) - 2\eta_{2BC} B^B V_D{}^j g^{CD}, \\
(M^{2A}{}_B &:= \delta^A{}_1 \delta^2{}_B + \delta^A{}_3 \delta^1{}_B) \\
\phi_i &:= R^1{}_{ij} V_3{}^j + R^2{}_{ij} V_1{}^j \\
\mathcal{H} &= 0, \; \mathcal{H}_i = 0, \; G_A = 0.
\end{aligned} \quad (3.4)$$

We note that $v_2{}^i \phi_i = \mathcal{H}$ and $V_1{}^i \phi_i = V_1{}^i \mathcal{H}_i$. Therefore, there are only 11 independent secondary constraints.

The triad rotations are generated by $-4i \int \Lambda^A G_A d^3x$, which has come to be called the Gauss-law constraint. From the Poisson brackets of $V_A{}^i$ with this constraint we find

$$\begin{aligned}
\delta_G V_1{}^j &= -2\Lambda^3 V_3^j + \Lambda^2 V_2{}^j \\
\delta_G V_2 &= -2\Lambda^1 V_2^j - \Lambda^3 V_1{}^j \\
\delta_G V_2 &= -2\Lambda^2 V_1^j - \Lambda^1 V_3{}^j
\end{aligned} \quad (3.5)$$

We require that $V_2{}^i$ remain zero. It follows that $\Lambda^3 = 0$ or rather that G_3 does not generate triad rotations which leave the null surface and its null generator invariant. This implies that G_3 cannot by a first class constraint.

We have altogether 24 primary and secondary constraints. One must now examine the constraint algebra and isolate the first class constraints from the second class. That is, one must find the maximum set of constraints whose Poisson brackets with all the other constraints vanish at least modulo the constraints themselves. This is a long and arduous task which I shall not present here. However, we shall give the results.

The 24 constraints separate into 12 first class and 12 second class constraints. Modulo the needed additions, the first class constraints are

$$\mathcal{H}_i = 0, \; G_{\mathbf{A}} = 0 \; (\mathbf{A} = 1, 2), \; P_{\mathbf{A}} = 0, \; p = 0, \; p_i = 0 \quad (3.6)$$

and the 12 second class constraints are

$$\mathcal{H} = 0, \; V_2{}^i = 0, \; \pi^2{}_i = 0, \; G_3 = 0, \; \chi^i = 0, \; V_3{}^i \phi_i = 0. \quad (3.7)$$

The 12 first class constraints imply conditions on 24 phase space variables, while the 12 second class constraints are only 12 such conditions. Thus altogether there are 36 conditions on the 38 phase space variables. This leaves two independent functions and corresponds to the one degree of freedom which exists on a characteristic surface.

4. Conlusion

This is only a sketch of the calculations and the results we have obtained. The details, including the Poisson bracket algebra of the first class constraints, are being prepared and will appear in a standard journal publication. Although the constraints look formidable, there are certain conclusions which we can draw. The second class constraints appear to impose conditions on $A^2 i$ and $v_2 i$. Coordinate conditions can be imposed so that $V_1 i$ has only one component; that is, the r direction is along the null geodesic of the surface. The null rotations can be used to limit the triad and with further gauge fixing of the first class constraints, only the complex shear of the null rays, $A^3{}_i v_2{}^i$ remains as the independent data. We have not yet examined this point, but the structure of the ϕ and χ equations give us hope that this can be accomplished explicitly.

At the same time, we wish to construct the Dirac brackets in order to eliminate the second class constraints without solving them explicitly. This may allow us to discuss the quantum theory of general relativity from this point of view.

5. Acknowlegment

We want to thank Abhay Ashtekar for several discussions of delicate points in this calculation. Most particularly, two of us (JNG and DCR) had in mind our relationship with Peter Bergmann from whom we learned so much about canonical theories with constraints. It is our pleasure to have been able to complete this much of our program to honor him on his 75th birthday.

The work of JNG was supported in part by the National Science Foundation under grant number PHY-9005790 and in part by a grant from the British Science and Engineering Research Council during the summer of 1989 when this work was begun. The work of DCR was supported in part by research funds from Syracuse University. CS is supported by funds from the British Science and Engineering Research Council.

Appendix

Lower case Greek indices will range from 0-3 while lower case Latin indices will range from 1-3. Letters from the beginning of the alphabet will label tetrads or triads. Letters from the middle or end of the alphabet will be coordinate indices.

Given a null tetrad

$$k^0 = Ndu, \qquad k^a = v^a{}_i(dx^i - N^i du),$$

$$ds^2 = 2k^0 k^1 - 2k^2 k^3, \qquad (A.1)$$

we construct three self-dual bivectors $\Sigma^{\mathbf{A}}(\mathbf{A} = 1, 2, 3)$

$$\Sigma^1 := \tfrac{1}{2}(k^1 \wedge k^0 + k^2 \wedge k^3) = \Sigma^{10} = \Sigma^{23},$$

$$\Sigma^2 := k^1 \wedge k^2 =: \Sigma^{12}, \qquad \Sigma^3 := k^3 \wedge k^0 =: \Sigma^{30}. \qquad (A.2)$$

We will use a "metric" on the upper case Latin indices defined by ($\epsilon^{\mu\nu\rho\sigma}$ is the Levi-Civita alternating tensor)

$$\Sigma^{\mathbf{A}}{}_{\mu\nu} \Sigma^{b\mathbf{B}}{}_{\rho\sigma} \epsilon^{\mu\nu\rho\sigma} = \frac{1}{3} g^{\mathbf{AB}} \Sigma^{\mathbf{C}}{}_{\mu\nu} \Sigma_{\mathbf{C}\rho\sigma} \epsilon^{\mu\nu\rho\sigma},$$

$$g^{\mathbf{AB}} = \begin{pmatrix} \tfrac{1}{2} & 0 & 0 \\ 0 & 0 & -1 \\ 0 & -1 & 0 \end{pmatrix} \qquad (A.3)$$

The self-dual components of the connection one-forms are defined by

$$\omega^+{}_{\alpha\beta} = \tfrac{1}{2}(\omega_{\alpha\beta} - \tfrac{1}{2} i \epsilon_{\alpha\beta\gamma\delta} \omega^{\gamma\delta})$$

$$dk^\alpha = \omega^\alpha{}_\beta \wedge k^\beta. \qquad (A.4)$$

We introduce a new symbol and labeling

$$\mathcal{G}^1 := -\omega^{+10} = -\omega^{+23}, \quad \mathcal{G}^2 := -\omega^{+12}, \quad \mathcal{G}^3 := -\omega^{+30}, \qquad (A.5)$$

so that

$$d\Sigma^{\mathbf{A}} = -2\eta^{\mathbf{A}}{}_{\mathbf{BC}},$$

$$\eta_{123} = -2\eta^{123} = 1. \qquad (A.6)$$

The Riemann tensor then has the definition

$$-\tfrac{1}{2} R^{\mathbf{A}} = d\mathcal{G}^{\mathbf{A}} + \eta^{\mathbf{A}}{}_{\mathbf{BC}}. \qquad (A.7)$$

Note that on the 4-space we may introduce a derivative operator on the bivector indices such that

$$\mathcal{D}_\rho \lambda_{\mathbf{A}} = \nabla_\rho \lambda_{\mathbf{A}} + 2\eta_{\mathbf{ABC}} \mathcal{G}^{\mathbf{B}}{}_\rho \lambda^{\mathbf{C}}. \qquad (A.8)$$

The pull-back of this derivative operator will be used on the null cone.

References

1. H. Bondi, M.G.J. van der Burg, and A.W.K. Metzner, Gravitational waves in general relativity: VII Waves from axi-symmetric isolated systems, Proc. Roy. Soc. **A269**, 21 (1962).

2. R.K. Sachs, Gravitational waves in general relativity: VIII Waves in asymptotically flat space-time, Proc. Roy. Soc. **A270**, 103 (1962).

3. E.T. Newman and T.W.J. Unti, Note on the dynamics of gravitational sources, J. Math. Phys. **6**, 1806 (1965).

4. J.N. Goldberg, The Hamiltonian of general relativity on a null surface, Found. of Phys. **14**, 1211 (1984).

5. J.N. Goldberg, Dirac brackets for general relativity on a null cone, Found. of Phys. **15**, 439 (1985).

6. C.G. Torre, Null surface geometrodynamics, Class. Quantum Grav. **3**, 773 (1986).

7. R.A. d'Inverno and J. Stachel, Conformal two-structure as the gravitational degrees of freedom in general relativity, J. Math. Phys. **19**, 2447 (1978).

8. R.A. d'Inverno and J. Smallwood, Covariant 2+2 formulation of the initial value problem in general relativity, Phys. Rev. **D22**, 1233 (1980).

9. A. Ashtekar, New variables for classical and quantum gravity, Phys. Rev. Lett. **57**, 2244 (1986).

10. J. Samuel, A Lagrangian basis for Ashtekar's reformulation of canonical gravity, Pramana J. Phys. **28**, L429 (1987).

11. T. Jacobson and L. Smolin, Covariant Action for Ashtekar's form of canonical gravity, Class. Quantum Grav. **5**, 583 (1988).

QUALITATIVE COSMOLOGY

I.M. Khalatnikov

L.D.Landau Institute for
Theoretical Physics
USSR Academy of Sciences

ABSTRACT

The article surveys qualitative methods of the study of the simplest cosmological models of the Universe in the dissipation-free regime and in the presence of viscosity. For homogeneous cosmological models the Einstein equations reduce to a system of regular differential equations with rspect to time. Further these equations are rewritten in the form of equations for a dynamical system permitting to analyze possible types of the evolution of solutions in the phase space of this system.

I. INTRODUCTION

The basis of the qualititive cosmology is the equations of the relativistic gravity theory, which actually are an extremely complicated system of differential nonlinear relations in terms of partial derivatives. For this reason in this field our studies are largely based on models. However only the simplest models are exactly integrable therefore even model investigations require this or that form of a qualitative analysis. Fortunately, most of the interesting cosmological models belong to the class of the so-called homogeneous models for which the equations of their dynamics reduce to systems of regular differential equations with respect to time. Consequently, such an efficient and well-developed method of study as the qualitative theory of dynamical systems can be employed in cosmology. Comparatively recently this theory has not been seriously applied to cosmological problems and only in the course of the last dozen years the situation has noticeably changed. First, methods of the qualitative theory were used in [1,2], where the problem reduces to dynamical systems on the plane. The theory of the oscillatory regime in the vicinity of the cosmological singularity [3,4] has inspired the development of the qualitative theory of multidimensional dynamical systems, emerging in homogeneous cosmological models of more sophisticated types [5,6]. Further, to study the problem of universality of such effects as the singularity in cosmology and the oscillatory regime in its vicinity, it was necessary to study the cosmological evolution of the matter with dissipative processes due to viscosity. It has been shown [7-10], that in this case also we can successfully and efficiently use methods of the qualitative theory to investigate possible types of the evolution in homogeneous cosmological models.

The results of this stage of developments of the qualitative methods in cosmology reveal that this approach to the study of homogeneous models is offen the most efficient approach, making it possible to comparatively easily obtain a comprehensive pattern of the dynamics of solutions there where an application of approximated analytical methods would bring about more complicated and less accurate results. In this article on the example of simplest homogeneous models I shall demonstrate the application of the qualitative theory of 2d dynamical systems to the study of the problem of the effect of dissipative effects on the character of the cosmological singularity and shall shown how easy it is to find an overall complexity of possible types of cosmological evolution of viscous liquid for the case under study by means of qualitative methods.

In Sect.II I shall describe the regular (dissipation-free) and well-developed isotropic Friedman models and the anisotropic type-I model in terms of the qualitative theory of dynamical systems. This makes it possible to introduce new notions on the basis of the already classical cosmological solutions and simplifies their comparison with solutions which emerge in the same models in the presence of dissipative processes. The qualitative pattern of this complexity of new solutions will be given in Sect.III of this article, where I shall also discuss the most interesting effects due to viscosity.

II. ANISOTROPIC COSMOLOGICAL TYPE-I MODEL AND CLASSICAL FRIEDMAN MODELS AS DYNAMICAL SYSTEMS ON THE PLANE

All homogeneous cosmological models can be of 9 types, depending on the symmetry properties of the respective 3d space (Biancchi classification). The simplest case is a type-I model where the 3d space in each given moment of time is flat (non-Euclidean). The 4d metric of the space-time for this model is written as[*]:

$$-ds^2 = -dt^2 + r_1^2(t)dx^2 + r_2^2(t)dy^2 + r_3^2(t)dz^2 \qquad (1)$$

Since the scale factors r_1, r_2 and r_3, describing the contraction and expansion of the space in the directions x,y,z are different in the general case, this model is anisotropic. The particular case when all the three scale coefficients are equal, describes the isotropic evolution and coincides with the case of the flat Friedman model.

The energy-momentum tensor of the perfect liquid filling the space is:

$$T_i^k = (\varepsilon+p)u_i u^k + p\delta_i^k, \qquad (2)$$

where ε and p are the energy density and the pressure related by a certain equation of state $p=p(\varepsilon)$.

A simple analysis of the Einstein equation

$$R_i^k = T_i^k - 1/2\delta_i^k T + \delta_i^k \Lambda \qquad (3)$$

(where for reasons of generality the Λ-term is also introduced) reveals

---

[*] A system of units where the velocity of light and the Einstein gravitation constant are unities. The metric is written in the form $-ds^2 = g_{ik} dx^i dx^k$, where g has a signature $(-+++)$. The Latin indices run over the values $(0,1,2,3)$, the Greek indices – $(1,2,3)$ and for the coordinates the designation $(x^0, x^1, x^2, x^3)=(t, x, y, z)$ is assumed.

that in the space with metric (1) the matter can be only comoving i.e., components of 4-velocity u^i must have the values

$$u^0 = 1, \quad u^\alpha = 0 \tag{4}$$

Then the 0α-components of the Einstein equations are satisfied identically and the unknown quantities which must be found from threomaining gravity equations, are the energy density $\varepsilon(t)$ and the three scale coefficients $r_\alpha(t)$. Here it is convenient to introduce quantities r and h in accordance with the following designations:

$$r_1 r_2 r_3 = r^3, \quad (\ln r)^{\cdot} = h \tag{5}$$

(here and elsewhere the dot $\cdot$ labels time-derivatives). Now taking the $\alpha\beta$-components of the gravity equations, it is easy to descover that at $\alpha \neq \beta$ they are satisfied identically and the remaining three diagonal components admit two first integrals, which can be written down symmetrically:

$$(\ln r_\alpha)^{\cdot} = h + s_\alpha r^{-3}, \quad \sum_\alpha s_\alpha = 0, \tag{6}$$

where s_α are the integration constants (only 2 of them are arbitrary). Then it is necessary to write out 00- and any of the $\alpha\alpha$-components of the equations. It is easy to show that with (6) these two equations reduce to:

$$\Lambda + \varepsilon = 3h^2 - l^2 r^{-b}, \tag{7}$$

$$\dot{h} - \Lambda + \varepsilon - 3h^2 - 1/2(\varepsilon+p), \tag{8}$$

where $l^2 = (s_1^2 + s_2^2 + s_3^2)/2$. The relations (5)-(8) constitute an overall set of the Einstein equations for the model under study. They, as is known, include the equations of hydrodynamics $T^k_{i;k} = 0$, which are quite handy for use. In this case the α-components of the hydrodynamical equations are satisfied identically and the 0-component yields

$$\dot{\varepsilon} = -3h(\varepsilon+p). \tag{9}$$

To define this model, one should also set an equation of state $p=p(\varepsilon)$. I shall confine myself to the most frequently employed in cosmology case of a linear pressure dependence on the energy density:

$$p = (\gamma - 1)\varepsilon, \tag{10}$$

where γ is a constant, confined in the limits:

$$1 \leq \gamma \leq 2 \tag{11}$$

It is not difficult to make sure that the equations written out in the general case connot be integrated exactly in terms of elementary functions. Yet, for many problems the exact solution is not needed. Very often we can regard the study of a cosmological model as completed if one can enumerate all permissible quantitatively possible types of solutions and also can give edge asymptotics, describing the initial and final stages of the evolution in each of these types. But the qualitative theory of dynamical systems is concered with such problems, i.e., with systems of regular first order differential equations to which Eqs.(5)-(9) belong. In terms of this theory enumeration of all possible types of cosmological solutions in this or that model reduces to definition of a qualitative structure of the phase space of the system under study, divided into cells, involving trajectories with some definite type of behaviour. The beginning and the termination of the evolution for each cosmological solution (i.e., for each trajectory) correspond to equilibrium states (or finite manifolds) of the respective dynamical system and a search for edge asymptotics reduces to one of the standard (not always simple) problems of the qualitative theory - to a search for an analytical form of the solution in the vicinity of these singularities or manifolds. These or those changes in the pattern of a cosmological regime induced by some changes of the external parameters of the model (e.g., constant Λ and γ) also affect the well-investigated field of qualitative methods - bifurcation theory [11]. Thus, the qualitative theory of dynamical systems proves to be quite well adapted to the analysis of homogeneous cosmological models and is the most adequate method of their investigation in cases when the exact solution is not possible.

Coming back to the type-I model notice, that at the given dependence $p(\varepsilon)$ Eqs. (8) and (9) single out from it a dynamical system in the phase space of variables ε and h. It appears that a similar system for analogous variables arises also in all the three versions of the Friedman model. Besides, the existence of a closed system in terms of these variables is retained also at generalizations of models under study by means of including the viscous terms into the energy-momentum tensor. For this reason (despite the fact that from Eqs.(5)-(8) one can single out also other closed systems) it is particularly interesting to study the plane (ε,h). So, Eqs.(8), (9) with (10) taken into account, become

$$\dot\varepsilon = -3\gamma\varepsilon h,$$

$$\dot h = \Lambda + \frac{2-\gamma}{2}\varepsilon - 3h^2 \qquad (12)$$

and describe a conservative (Hamiltonian) dynamical system, admitting an exact integral

$$\left[3h^2 - (\Lambda+\varepsilon)\right]\varepsilon^{-2/\gamma} = C, \qquad (13)$$

where C is an arbitrary constant. It is easy to see that this constant cannot be negative. Actually, the relation (7) and the requirement that the energy density should be positive, set additional constraints for the system (12), selecting as physical only trajectories lying in the domain[*)]

$$\varepsilon \geq 0, 3h^2-(\Lambda+\varepsilon)\geq 0. \qquad (14)$$

Hence, all trajectories, corresponding to the negative values of C, lie in a nonphysical region of the phase plane: inside the parabola $\varepsilon=3h^2-\Lambda$.

The phase pattern of the dynamical system (12) under the additional constraints (14) and for all possible values of the sign of the cosmological constant is given in Fig.1. The trajectories in the physical part of the phase plane (ε,h) are depicted in the upper part of Fig.1. Under each of the upper schemes there is its respective orthogonal projection onto the plane (ε,h) of the physical part of the lower half of

---

[*] Then in Expr.(13) here and hereafter ε^α (where α is an arbitrary number) is regarded as an arithmetic value of this quantity. The same concerns all powers of positive numbers we shall come across hereafter.

the Poincare sphere[*]) (for $\gamma \neq 1$), which enables us to see the character of these trajectories on the infinity. In the upper part of the plane (ε,h), where $h>0$, there are trajectories describing the expansion of the space, in the lower half plane $(h<0)$ - the contraction. The boundaries of the physical region are also integral curves of Eqs.(12): on the axis $\varepsilon=0$ there are trajctories of vacuum solutions, on the parabola $\varepsilon=3h^2-\Lambda$ - trajectories of the isotropic flat Friedman model. The arrows on the trajectories correspond to the increase of the physical time t.

Consider a case without the cosmological constant $(\Lambda=0)$, shown in Fig.1,b. We see that the dynamical system (12) in this case has four singular points: one in the origin of the phase plane and three - on the infinity. The points 0 and I are complex equilibrium states, having in the physical region one stable and one unstable knot sectors, the points I_1 and I_2 are saddles. All trajectories $I_o 0$ and one vacuum trajectory $I_1 0$, outgoing from the infinity to the origin, describe the expansion of the space from the cosmological singularity of the "Big Bang" type to the final stage of an infinite expansion in the point 0, where ε and h tend to zero. These evolutions, as is easy to show, correspond to the change of time t from a certain finite initial value $t=0$ up to $t=+\infty$. In the lower

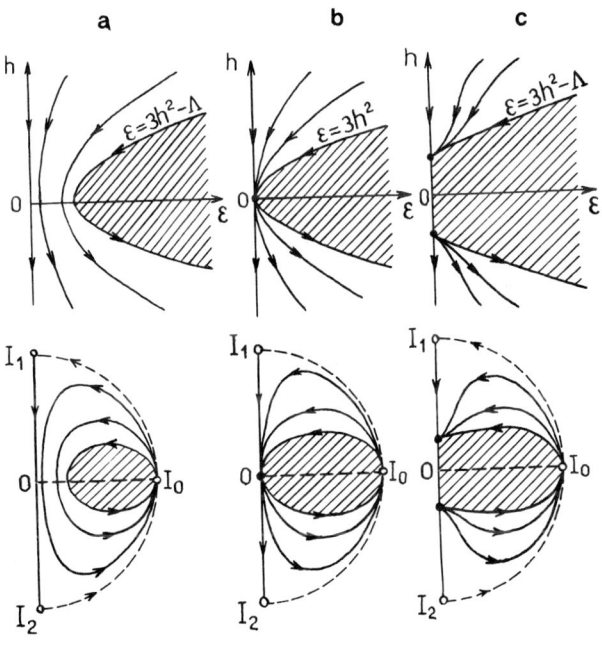

Figure 1

---

[*])The sphere of a unit radius lying on the phase plane (if to represent the latter in the horizontal position) and tangent to it in the origin. All points of the phase plane are mapped onto the lower half-sphere by means of the central projection. Then the equator is the mapping of infinitely remote points of the plane. Then again mapping the lower half-sphere onto the phase plane, but already by means of the orthogonal projection, we get a topological mapping of the whole phase space onto the inside of the circle of a unit radius. The character of the points of the boundary of the circle and the behaviour of the trajectories near this boundary completely describe properties of the phase infinity [12].

outgoing from the infinity to the origin, describe the expansion of the space from the cosmological singularity of the "Big Bang" type to the final stage of an infinite expansion in the point 0, where ε and h tend to zero. These evolutions, as is easy to show, correspond to the change of time t from a certain finite initial value t=0 up to t=+∞. In the lower half-plane (h<0) all trajectories $0I$ describe the contraction of the space from the infinitely expanded stage to the state of collapse where the energy density and the quantity h turn into infinity. The vacuum trajectory $0I_2$ describes an analogous collapse of an empty space. All these evolutions correspond to the change of time from t=-∞ up to a certain finite moment, which could be t=0. Moreover, the invariance of Eqs.(12) and of the constraints (14) with respect to the substitution

$$t \to -t, \quad h \to -h \tag{15}$$

implies that all solutions of the lower half-plane, corresponding to the contraction, can be obtained by the substitution (15) from the solutions represented by the trajectories of the upper half-plane and describing the expansion of the model. Due to this symmetry of the phase plane it is sufficient to analyse only the trajectories $I_0 0$ and $I_1 0$, corresponding to the positive values of t from the interval [0, +∞]. Before giving the edge asymptotics for this set of trajectories, note that its boundaries (i.e., vacuum and Friedman cases) correspond to solutions admitting exact elementary integration. The trajectory $I\,0(\varepsilon \equiv 0)$ corresponds to the well-known Kasner solution:

$$-ds^2 = -dt^2 + t^{2p_1}dx^2 + t^{2p_2}dy^2 + t^{2p_3}dz^2, \tag{16}$$

$$p_1 + p_2 + p_3 = 1, \quad p_1^2 + p_2^2 + p_3^2 = 1, \tag{17}$$

$$h = (3t)^{-1}, \tag{18}$$

and the trajectory $I_0 0$, going along the parabola $\varepsilon = 3h^2$, corresponds to the flat Friedman model:*)

$$-ds^2 = -dt^2 + t^{4/3\gamma}(dx^2+dy^2+dz^2), \tag{19}$$

$$\varepsilon = 4/3\gamma^2 t^2, \quad h = 2/3\gamma t. \tag{20}$$

The values C=∞ and C=0 in the integral (13) correspond to the written exact solutions. The intermediate values of the constant C correspond to all other trajectories $\gamma \neq 2$ and it is easy to show that at $\gamma \neq 2$ the asymptotic form of the metric for these solutions in the vicinity of the initial singularity t=0 irrespective of the presence of the matter, has the same vacuum form (16)-(18). The asymptotic behaviour of the energy density at t→0 for such solution is

---

* In Exprs. (16) and (19) as well as in all similar expressions for the metric, we shall come across further, for simplicity, we shall not write arbitrary constant factors which can be eliminated by the scale transformation of the coordinates x,y,z.

$$\varepsilon = \left(\sqrt{3C}\, t\right)^{-\gamma}. \tag{21}$$

all intermediate trajectories I_oO are terminate by the stage of an infinite expansion in the point 0, where $\varepsilon \to 0$. In this region they all, as is evident from the integral (13) at $\gamma \neq 2$, merge with the Friedman parabola $\varepsilon = 3h^2$. It means that all these cosmological evolutions at $t \to \infty$ have the asymptotics (19)-(20), i.e. at the expansion the model sooner or later isotropizes.

The both effects, i.e., the negligibly small influence of the matter near the "Big Bang" moment and isotropization of the regime on the final stages of the expansion, do not occur at $\gamma = 2$ (for the "stiff" equation of state $p = \varepsilon$). In this case the equations of the investigated model can be integrated exactly and yield the following solution:[*]

$$-ds^2 = -dt^2 + t^{2q_1}dx^2 + t^{2q_2}dy^2 + t^{2q_3}dz^2 \tag{22}$$

$$\gamma = 2, \quad \varepsilon = \left[3(1+C)t^2\right]^{-1}, \quad h = (3t)^{-1}, \tag{23}$$

$$q_1 + q_2 + q_3 = 1, \quad q_1^2 + q_2^2 + q_3^2 = 1 - 2\left[3(1+C)\right]^{-1} \tag{24}$$

(at $C \to \infty$ this solution turns into an exact vacuum solution and at $C \to 0$ - into an exact Friedman solution).

Despite the fact that the value of the parameter $\gamma = 2$ is exceptional, the qualitative behaviour of the trajectories, given in Fig. 1,b, does not alter at the transition from the values $\gamma < 2$ and $\gamma = 2$, i.e., there are no bifurcations[**]. The lower boundary $\gamma = 1$ of the range of variation of the parameter γ, dividing the regions of positive and negative values of the pressure p, is also physically exceptional and here actually bifurcations do occur but only on the infinity. It is clear from (13) that at $\gamma = 1$ the equations of the trajectories at $\varepsilon \to \infty$ transform into the equations of the straight lines $h = \pm\sqrt{C/3}\,\varepsilon$, and their infinitely remote edges are not already mapped onto one point on the equator of the Poincare sphere. Thus, the saddle sectors of the point I_o vanish and the phase pattern of the system corresponds to the one given in Fig.2. This case ($\Lambda = 0$, $\gamma = 1$) also belougs to the exactly integrable class. The

---

[*] This solution has a certain particular place in cosmology. It is possible to show [13.14], that at the equation of state $p = \varepsilon$ even in the general cosmological solution of the Einstein equations when the singular point is being approached, the oscillatory regime sooner or later ceases and in the vicinity of the singularity there is a power asymptotics of the (22)-(24) type, corresponding to the case when all the three factors q_α are positive.

[**] In reality when the value $\gamma = 2$ is achieved, in the dynamical system there occur bifurcations (of different character at different values of Λ), but they occur in the nonphysical region of the phase plane either at $\varepsilon < 0$ or at $3h - (\Lambda + \varepsilon) < 0$.

appropriate solution was first derived by O. Heckman and E.Schuking in 1958. At small t its asymptotics is given by Formulas (16)-(18) and (21) and at t→∞ - by the Friedman solution (19)-(20).

Qualitative changes occuring at the variation of the parameter γ in the physical region are not important in comparison with those which emerge when the non-zero cosmological constant Λ arises. A simple analysis of the system (12) shows that the emergence of a negligibly small $\Lambda>0$ is accompanied by the decay of the complex equilibrium state 0 into three simple states: one saddle and two knots. The saddle point is in the region $\varepsilon<0$ and is of no interest whereas the simple knots K_1 and K_2 emerge in the points of the intersection of the Friedman parabola $\varepsilon=3h^2-\Lambda$ with the vacuum axis $\varepsilon\equiv0$. The respective phase pattern is given in Fig.1,c. At the change of Λ from 0 towards negative values the complex singular point transforms into one simple point (center or focus), lying in the nonphysical region $3h-(\Lambda+\varepsilon)<0$. In the physical region of the phase plane there are no equilibrium states left (except on the infinity, naturally). The pattern of trajectories for this case is given in Fig.1,a. Thus, the zero value of the cosmological constant is bifurcational. Qualitatively new phenomena at $\Lambda\neq0$ arise, however, only in the finite vicinity of the origin of the phase plane, do not affect the infinity and do not alter the structure of the system (12) on the equator of the Poincare sphere. It means that the influence of the Λ-term in the vinicity of cosmological singularities I_0, I_1, I_2 is negligibly low and the asymptotics of the solutions in these regions is identical to the one in the case of the absence of the cosmological constant (the only defference being that at $\Lambda\neq0$ Exprs.(16)-(20) and (22)-(24) yield only at t→0 an asymptotics but not exact solutions). The phenomenon of bifurcation of the infinity at $\gamma=1$ shown in Fig.2, is not affected either.

The final stages of the expansion of the model now have an essentially defferent character. At $\Lambda<0$ there are no final stages at all. Any trajectory (even vacuum) describes a closed cycle of evolution, consisting of the "original explosion", expansion maximum and consecutive contraction down to the singularity. In the course of this evolution the time changes in a limited interval, increasing from t=t (initial singularity) up to t=t$_2$ (collapse).[*]

At $\Lambda>0$ the trajectories in the upper and lower half-planes of the phase space, as usual, belong to independent classes, obtained from one another by means of the transformation (15), therefore it is sufficient to consider only the solutions corresponding to the expansion. As is shown in Fig.1,c at the final stages of the expansion all trajectories get into the knot K_1. This equilibrium state has the coordinates $\varepsilon=0$ and

---

[*] At $\Lambda<0$ the trajectories in the upper and lower half-planes do not already belong to two independent classes. Therefore the same value of time t=0 (but from different sides of this point) cannot formally be correlated with the beginning and the end of the evolution. In this connection the asymptotics in the vicinity of the original singularity now ensue from Formulas (16) - (24) where one should perform the substitution t→t-t$_1$, and the asymptotics near the collapse is obtained from the same formulas by the substitution h→-h and t→t$_2$-t. It should also be from in mind that now by the scale transformations of the coordinates x,y,z one can eliminate arbitrary factors in the metric coeffecients r_α only in the vicinity of one singularity (either at t=t$_1$, or t=t$_2$).

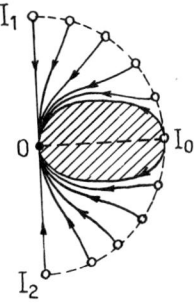

Figure 2

$$h = \sqrt{\Lambda/3} \qquad (25)$$

and corresponds to the exact vacuum De Sitter solution:

$$-ds^2 = -dt^2 + e^{2\sqrt{\Lambda/3}\,t}(dx^2+dy^2+dz^2). \qquad (26)$$

All the trajectories tend to the knot K_1 at $t \to +\infty$ exponentially fast. The energy density then decreases according to the law

$$\varepsilon = \text{const} \cdot e^{-\gamma\sqrt{3\Lambda}\,t}. \qquad (27)$$

Formulas (25)-(27) describe the asymptotics of the final stages of the expansion at $\Lambda > 0$. Then all solutions (including the vacuum solution and those corresponding to $\gamma=2$) at $t \to \infty$ isotropize.

Now consider isotropic cosmological Friedman models by means of the qualitative methods [8,9]. The metrics of these models can be conveniently written down as

$$-ds^2 = -dt^2 + \frac{R^2(t)\,(dx^2+dy^2+dz^2)}{[1+k\,(x^2+y^2+z^2)/4]^4}, \qquad (28)$$

where the cases $k=+1$, $k=-1$ and $k=0$ correspond to the closed, open and flat 3d space, respectively. If to introduce the Habble "constant"

$$H = (\ln R)\dot{}, \qquad (29)$$

it is easy to show that the Einstein equations (3) with the energy-momentum tensor (2) in the respective system (4) as well as the equations of hydrodynamics are written down in the form of the three relations

$$\Lambda + \varepsilon = 3H^2 + 3kR^{-2}, \qquad (30)$$

$$\dot{H} = 1/2(\Lambda + \varepsilon - 3H^2) - 1/2(\varepsilon + p), \qquad (31)$$

$$\dot{\varepsilon} = -3H(\varepsilon + p). \qquad (32)$$

For the equation of state (10), from (31) and (32) we obtain a dynamical system in the phase plane (ε, H) analogous to the system (12)

$$\dot{\varepsilon} = -3\gamma\varepsilon H,$$

$$\dot{H} = \Lambda/3 + \frac{2-3\gamma}{6}\varepsilon - H^2. \qquad (33)$$

This system is also conservative and admits an exact integral for phase trajectories, analogous to the integral (13)

$$[3H^2 - (\Lambda+\varepsilon)]\varepsilon^{-2/3\gamma} = C_1. \qquad (34)$$

Yet, now the additional conditions for Eqs.(33), singling out the physical region of the plane (ε, H) are considerably altered. We see that Eqs.(33) do not explicitly involve the parameter k and, consequently, describe on the same sample of the phase plane the trajectories belonging to all the three types of Friedman models simultaneously. The relations (30) then show that the trajectories of the flat model (k=0), as previously, lie on the parabola $3H^2=\Lambda+\varepsilon$, and the solution for the open model (k=-1) as previously get into a region beyond this parabola, where $3H^2-(\Lambda+\varepsilon)>0$. The inside of the parabola $3H^2-(\Lambda+\varepsilon)<0$ is now a physical region and is filled with the trajectories of the closed model (k=+1). Correspondingly, the arbitrary constant C_1 in (34) takes over the values $C_1=0$ for the flat, $C_1>0$ for the open and $C_1<0$ for the closed space. The pattern of phase trajectories of the system (33) for defferent values of Λ is shown in Fig.3.

In Figs.3,b describing the case $\Lambda=0$, we see the already familiar families of trajectories for the open space to which integral curves inside the parabola $\varepsilon=3H^2$ describing the entire cycle of the evolution of the closed Friedman model are added. For all trajectories, except the vacuum one, the asymptotics near the moment of the original singularity $t=t_1$ (exit from the infinity I_o) has the form

$$R = R_1(t-t_1)^{2/3\gamma}, \quad \varepsilon = 4/3\gamma^2(t-t_1)^2,$$

$$H = 2/3\gamma(t-t_1), \qquad (35)$$

and near the moment of the collapse $t=t_2$ (entrance into I) it is written as

$$R = R_2(t_2-t)^{2/2\gamma}, \quad \varepsilon = 4/3\gamma^2(t_2-t)^2,$$

$$H = 2/3\gamma(t-t_2), \qquad (36)$$

where R_1, R_2 are arbitrary constants. For open models one can choose $t_1=t_2=0$ and correlate the region of time variation $[0, +\infty]$ with the expansion and the region $[-\infty, 0]$ - with the contraction. At such a choice the solution for the vacuum trajectory $I_1 0(\varepsilon\equiv)$, which proves to be exact, is written down as

$$R = t, \quad H = 1/t, \qquad (37)$$

and the solution for the trajectory $0I_2$ is obtained by the substitution

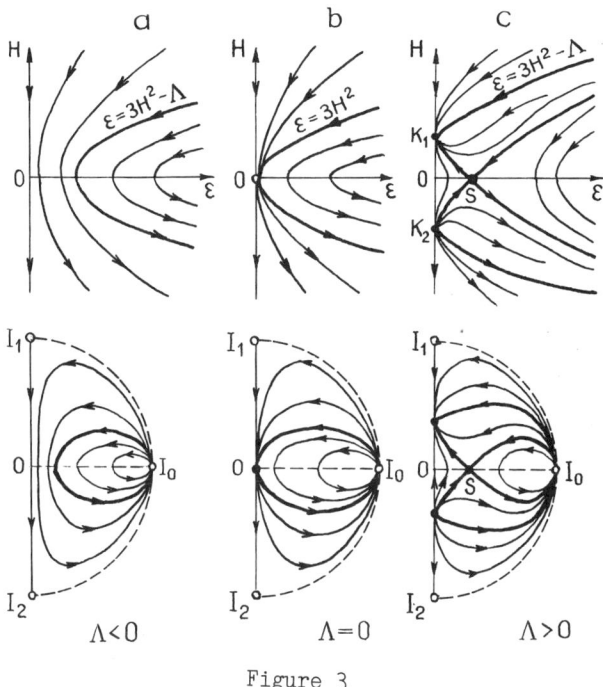

Figure 3

$$t \to -t, \quad H \to -H. \tag{38}$$

These vacuum solutions in the case $\Lambda=0$ yield, as is known, the flat Minkowski space-time.

The asymptotics of the solutions on the final stages of the expansion in the open model is clear from (34). Near the point $0(t\to+\infty)$ the energy density tends to zero and it follows from (34) that all the trajectories merge with the vacuum trajectory $I_1 0$. Thus, the asymptotics of the metric in the open model at $t\to\infty$ is given by Formulas (37) and the energy density behaves according to the law

$$\varepsilon = (3/C_1)^{3\gamma/2} t^{-3\gamma}. \tag{39}$$

For the flat model $C = 0$ the evolution is also described by the exact solution (19)-(20) (where h should be replaced by H).

The dependence of the character of solutions in isotropic models on the parameter γ is weaker than the one in the anisotropic model of type I. For Friedman solutions the values $\gamma=1$ and $\gamma=2$ are not already preferable. So, for instance, the bifurcation of the infinity at $\gamma=1$ is absent. As for bifurcations over the cosmological constant, they remain as they were. At $\Lambda<0$ in the physical region of the phase plane the equilibrium state at finite values of ε and H vanishes and the phase pattern of the system (33) acquires the form depicted in Fig.3,a. Now all trajectories behave like in the closed model: each evolution has its expansion maximum followed by the phase of contraction. The asymptotics on the infinity I_0 are given by the same formulas (35) and (36) as at $\Lambda=0$. The edge asymptotics of the vacuum trajectory $I_1 I_2 (\varepsilon=0)$ are obtained from (37) like it has been explained in the footnote on page 13. Near the beginning of the evolution $I_1(t=t_1)$ we

have
$$R = r_1(r-t_1). \quad H = 1/(t-t_1), \qquad (40)$$

and near its end $I_2(t=t_2)$ one can write

$$R = R_2(t_2-t), \quad H = 1/(t-t_2). \qquad (41)$$

At $\Lambda \neq 0$ the vacuum solution $I_1 I_2$ actually discribes the curved space-time but its edge singularities nevertheless are non-physical and are eliminated by the coordinate transformation of the same form by which the metrics (28) at $k=-1$ and $R=t$ are reduced to the Minkowski metric.

The appearance in the system (33) of a negligibly small positive constant Λ leads, like in the anisotropic case, to the decay of the complex equilibrium state 0 into two simple knots K_1, K_2 and one saddle S. Now, however, the saddle gets into the physical region and has the coordinates

$$\varepsilon = 2\Lambda/(3\gamma-2), \quad H = 0. \qquad (42)$$

The knot points K_1 and K_2, as previously lie on the axis $\varepsilon=0$ at the following values of the Habble "constant"

$$H = \sqrt{\Lambda/3} \text{ and } H = -\sqrt{\Lambda/3} \qquad (43)$$

respectively. The behaviour of the trajectories for the case $\Lambda>0$ is given in Fig.3,c.

Like in the already studied cases, the influence of the Λ-term on the infinity of the phase plane is negligibly low and the asymptotics of the solutions near the points I_0, I_1, I_3 are not distinct from those available for the trajectorics at $\Lambda=0$. In the vicinity of the origin in the open and flat models in comparison with the case $\Lambda=0$, there occur the same changes as in the anisotropic type-I model: the final stages of the expansion terminate in the knot K_1 at $t \to +\infty$ with the same De Sitter asymptotics (26)-(27).

Essentially new types of cosmological evolutions arise at $\Lambda>0$ for the closed Friedman model. Above all, this is a saddle point S which is the same stationary Einstein Universe, for the sake of which the Λ-term was introduced into the gravity equation. The presence of this saddle leads to the existence of the triangle $K_1 S K_2$ with trajectories at $t=-\infty$ outgoing from the knot K_2 and at $t \to +\infty$ incoming into the knot K_1. These solutions begin and terminate with the De Sitter metrics and nowhere in time have any physical singularities. In the triangle $K_1 I_0 S$ there are trajectories representing evolutions of the closed model but infinitely expanding without the phase of contraction. The final stage of this expansion is the same De Sitter asymptotics in the point K_1. The solutions are represented by integral curves of the triangle $K_2 I_0 S$. And finally, the region $SI_0 S$ contains the regular solutions, describing a closed cycle of the evolution of the closed model.

3. DYNAMICS OF COSMOLOGICAL MODELS IN THE PRESENCE OF VISCOSITY

In the preceding sections we have carried out a study of homogeneous models, describing the cosmological evolution of a perfect liquid with the

In the preceding sections we have carried out a study of homogeneous models, describing the cosmological evolution of a perfect liquid with the energy-momentum tensor (2). One of the fundamental properties of the studied solutions at $\Lambda=0$ [*]) is an obligatory presence in them of a cosmological singularity, marking the beginning (or the end) of the evolution of the Universe. One can show that the emergence of this singularity is inherent not only in homogeneous models but also in the general cosmological solution of the equations of the relativistic gravitation theory although in the general inhomogeneous case the behaviour of the gravity field and matter in the vicinity of this singularity is characterized by a sophisticated oscillatory regime [3,4]. Of interest is the investigation of the degree of universality of a phenomenon like an cosmological singularity not only with respect to generalizations for the case of inhomogeneous space but also with respect to generalizations of the form of the energy-momentum tensor taking into account a more real description of the behaviour of the matter. One of these generalizations is associated with the account of dissipative processes due to viscosity and it is natural to begin the appropriate analysis again with the siumplest homogeneous models described in the preceding sections. Already on these examples one manages to draw a number of important conclusions which are also valid in the general inhomogeneous case. Above all it becomes clear that the action of viscosity cannot eliminate the cosmological singularity although it introduces qualitatively new elements into the character of solutions. In the presence of dissipative processes the cosmological evolution becomes irreversible in time (the symmetry of (15) and (38) vanishes), which brings about differences in the patterns of expansion and contraction of the Universe. The general qualitative character of expansion is invaried: the collapse starts from the state of maximal expansion (finite or infinite) and terminates with the singularity at a certain finite moment of eigentime accompanied by the fact that the energy density and invariants of the curvature turn into infinity.

For the process of the expansion of the space there is a richer variety of possibilities. Above all, there emerges a diversity of solutions with a principally new character of the initial stage. In the vicinity of the initial stage the influence of the matter is negligibly low and the original singularity is determined by the Einstein equations in vacuum. Then, which is more important the energy density of the matter in the moment of the beginning turns into zero, then increasing in the course of the consecutive expansion. Thus, there arises a model of the evolution with an interesting property: in the course of the emergence of the Universe the matter is being created. There appear possibilities also for the final stages of the expansion. Apart from the solutions with the conventional Friedman infinite expansion at $t\to\infty$ there also emerge solutions having at $t\to\infty$ the De Sitter asymptotics despite the fact that there is no Λ-term in the equations (the effective cosmological constant is created due to the dissipation effect). Of interest is also a new type of evolutions with the beginning involving the above mentioned matter creation effect, terminating with the conventional Friedman expansion at $t\to\infty$ but inbetween these stages having a time limited stage close to the exponentially fast De Sitter expansion. The appearance in the process of

---

[*]) Hereafter we shall confine ourselves to the analysis of only conventional gravity equations without the cosmological term. This is also justified by the fact that the most important dissipative effects in the

the evolution of the De Sitter stage, then replaced by the Friedman expansion, at present attracts a great attention since it gives new possibilities to account for a number of observational data in this model. And finally, the quite natural process of the entropy increase in the course of the evolution of models with viscosity can also be referred to as a satisfactory property of such models. This property also models a certain aspect of the reality since it is known that the specific entropy in the modern Universe is rather high.

Now in more detail consider qualitative properties of models with viscosity and show how the above described effects occur. The energy-momentum tensor of a viscous liquid is written out as [15]

$$T_{ik} = (\varepsilon+p')u_i u_k + p'g_{ik} -$$
$$- \eta(u_{i;k} + u_{k;i} + u^k u^l u_{i;l} + u_i u^l u^k_{;l}),$$
$$p' = p - (\zeta-2\eta/3)u^k_{;k}, \quad u^k u_k = -1. \tag{44}$$

Here $\eta > 0$ and $\zeta > 0$ are the coefficients of the first (shear) and second (bulk) viscosities, respectively. These coefficients, like all other scalar quantities (e.g., ε and p) in homogeneous models are only time dependent and therefore we can treat them as functions of the same variable, i.e., of the energy density ε. For isotropic Friedman models Exprs.(44) are of the simplest form. Herefore we shall first consider this case. Actually, at the isotropic cosmological evolution there are no displacement of the layers of the matter with respect to each other, in virtue of which the shear viscosity does not in any way manifest itself and the gravity equations will be identical to those for the case when the coefficient of this viscosity η is zero. Taking into account only the bulk viscosity, we find that for the metric (28) under the condition (4) energy-momentum tensor will read

$$T^k_i = (\varepsilon+p')u_i u^k + p'\delta^k_i, \quad p' = p-3\zeta H, \tag{45}$$

where H is the same Habble "constant" (29). Now the equations of hydrodynamics $T^k_{i;k} = 0$ and the Einstein equations (3) at $\Lambda=0$ reduce to the three relations:

$$\varepsilon = 3H^2 + 3kR^{-2}, \tag{46}$$

$$\dot H = (\varepsilon-3H^2)/3 - (\varepsilon+p)/3 + 3\zeta H/2, \tag{47}$$

$$\dot\varepsilon = -3H(\varepsilon+p) + 9\zeta H^2, \tag{48}$$

from which it is clear that at the fixed dependences $p=p(\varepsilon)$ and $\zeta=\zeta(\varepsilon)$ Eqs.(47) and (48), as previously, describe the dynamical system in the phase plane (ε,H). Take the equation of state in the same form as (10), which brings us to the system

$$\dot\varepsilon = 3H(3\zeta H - \gamma\varepsilon)$$

$$\dot{H} = (\varepsilon - 3H^2)/3 + (3\zeta H - \gamma\varepsilon)/2. \tag{49}$$

As for the viscosity coefficient $\zeta(\varepsilon)$ first of all we need to know its asymptotic behaviour in the regions of small and large values of the energy density. The analysis of a relativistic kinetic equation for certain simple cases[*] as well as general physical considerations enable us to assume that in these regions the coefficient ζ can be approximated by power functions if some requirements are applied to power indices. At small values of ε it is reasonable to regard this ex[pmemt as larger than unity (or at least equal to unity) and in the region of large values of ε this exponent must be smaller than 1/2:

$$\zeta = \zeta_1 \varepsilon^{\nu_1}, \ \nu_1 \geq 1 \ (\text{at } \varepsilon \to 0)$$

$$\zeta = \zeta_2 \varepsilon^{\nu_2}, \ 0 < \nu_2 < 1/2 \ (\text{at } \varepsilon \to \infty), \tag{50}$$

where ζ and ζ are arbitrary positive constants.

The asymptotics (50) provide the principally necessary information for a complete analysis of the dynamical system (49). Above all, we shall also require that the function $\zeta(\varepsilon)$ between the points $\varepsilon=0$ and $\varepsilon=\infty$ should have no zeroes and infinities and should be reasonably smooth. Now, making use of Exprs.(50) we can find a qualitative pattern of the trajectories of the system (49) in the boundary regions of the phase plane $\varepsilon \to 0$ and $\varepsilon \to \infty$ and find the edge asymptotics of the solutions (the appropriate research was in detail covered in [8]). To merge the obtained edge parts of the phase pattern one needs to know the position and the character of equilibrium states of the system (49) inbetweeen these[**] regions, i.e., at finite and non-zero values of ε and finite values of H. From Eqs.(49) it is clear

that in the presence of viscosity such singular points (previously absent) may actually arise up. If they exist, they can be located only on the upper branch (H>0) of the parabola $\varepsilon = 3H^2$ and its coordinates can be found from the relations:

$$\zeta(\varepsilon) = \gamma\sqrt{\varepsilon/3}, \ H = \sqrt{\varepsilon/3}. \tag{51}$$

Thus the presence of these equilibrium states is determined by intersection or tangences of the curves $\zeta(\varepsilon)$ and $\gamma \sqrt{\varepsilon/3}$. It is easy to see that under the condition (50) these curves either do not intersect at all or have an even number of intersections. It is not difficult to find that in the latter case the respective equilibrium states are simple and have the character of either a saddle or of a simple stable knot. Then the first of these singular points (if to go along the parabola $\varepsilon = 3H^2$ from the origin) is necessarily a saddle point and the next one is a knot point.

---

[*] The appropriate estimation has been performed by A.Starobinsky.

[**] It is easy to see that under the conditions of the behaviour of the function $\zeta(\varepsilon)$ we have taken, the equilibrium states on the infinity with respect to H cannot arise in this intermediate region.

Further all of them alternate according to the same rule-saddle-knot-saddle-knot... The creation of such points can be regarded as bifurcations induced by the variation of the form of the function $\zeta(\varepsilon)$ towards its complication. Fig.4 gives the beginning of this process: the transition from the case without singular points (a) to the case when there are two singular points (c). The intermediate stage (b) corresponds to the case when the curves $\gamma(\varepsilon)$ and $\gamma\sqrt{\varepsilon/3}$ are tangent and, as the analysis reveals, yields a complex equilibrium state of the saddle-knot type. At the transition from (b) to (c) this complex point desomtegrates into 2 simple points: saddle S and knot K. Further complication of the curve $\zeta(\varepsilon)$ may generate new complex equilibria and their decay into simple points.

Nevertheless it is important that without restricting the generality of our research, we can confine ourselves to the study of only those principal cases which correspond to Fig.4. Actually it is not difficult to make sure that in the presence of more than two singular points the phase plane is divided by the separatrices of the saddles into qualitatively identical cells and the in crease of the number of equilibrium states does not generate qualitatively new types of solutions. For a similar reason we can also neglect the version with a complex singular point, corresponding to Fig.4,b since all types of the solutions emerging here are contained in the case with two singular points. Thus, there are only 2 main possible behaviours of the function $\zeta(\varepsilon)$ between the boundaries $\varepsilon=0$ and $\varepsilon=\infty$: without common points with the eurve $\gamma\sqrt{\varepsilon/3}$ (Fig.4,a) and with double intersection with this curve (Fig.4,c).

The phase patterns of the dynamical system (49) for these main cases are depicted in Fig.5. In the case represented by Fig.5,a (if to compare it with the case of a perfect liquid shown in Fig.3,b) we observe bifurcation on the infinity of the phase plane originating from the appearance of dissipation in the model under study. The point I_1 turns into a saddle, in the sector $I_1 I_0$ there emerges a separatrix OL and part of the trajectories start to go out from the infinity $(\varepsilon,H)=(0,+\infty)$. The asymptotics of the solution near the origin 0 are the same as in the case without viscosity: the incoming trajectories $(t\to+\infty)$ describe purely Friedman stages of unlimited expansion, the outgoing trajectories $(t=-\infty)$ - the beginning of the contraction. The initial and final singularities in the point I_o for all the trajectories, except the separatrix OL, are described by the same asymptotic formulas as in the absence of viscosity. The separatrix OL is the single solution where the influence of viscosity manifests itself also in the asymptotic vicinity to the initial singularity. Yet, his influence under the conditions (50) has only a quantitative character and, from the qualitative point of view, is irrelevant. Thus, an essentially new effect is the emergence of the trajectories I 0. The point I corresponds to the finite moment of time $t=0$ and near it the influence of viscous terms in Eqs.(49) proves to be dominating $(3\zeta H > \gamma\varepsilon)$. In the beginnig of such evolutions the energy density turns into zero and then increases in the course of the expansion of the space up to a certain maximum and after that again falls down to zero at $t\to\infty$ in conformity with the Friedman laws of the expansion on these final stages. These solutions, in fact, yield the model of creation of the matter in the beginning of the cosmological evolution, we have already spoken about previously. The asymptotic form of the solutions near I $(t\to 0$ and $t>0)$ depends on the exponent ν_1. If $\nu_1 > 1$, the asymptotics has a power-like form

$$R = t, \quad H = 1/t, \quad \varepsilon = \left(\frac{1+3\gamma(\nu-1)}{9\zeta_1(\nu_1-1)} t\right)^{1/(\nu_1-1)}. \qquad (52)$$

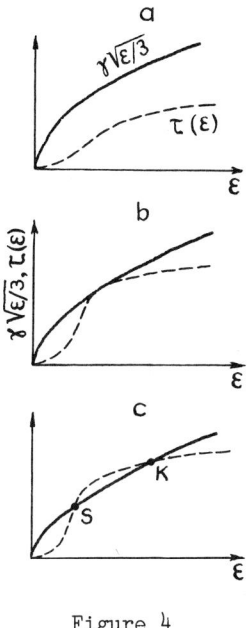

Figure 4

It is easy to see that here the quantities $3\zeta H$ and $\gamma\varepsilon$ are of the same order in time but nevertheless

$$\frac{3\zeta H}{\gamma\varepsilon} = \frac{1+3\gamma(\nu_1-1)}{3\gamma(\nu_1-1)} > 1.$$

If $\nu_1=1$, Formulas (52) are not applicable. The asymptotics of the solution in this case has an essentially different character:

$$R = t, \quad H = 1/t, \quad \varepsilon = \text{const} \cdot e^{-9\zeta_1/t} \quad (\nu_1 = 1) \tag{53}$$

Then in the region $t \to 0$ $3\zeta H \gg \gamma\varepsilon$.

Fig.5,b depicts the case with two simple equilibrium states on the upper branch of the parabola $\varepsilon = 3H^2$. In comparison with the preceding case this one has new possibilities for the final stages of the expansion at $t \to \infty$. In all the three types of Friedman models there are now trajectories caused by the "Big Bang" in the point I_o (near it the influence of viscosity is, as previously, negligibly low) and terminating at $t \to \infty$ in the knot K with the exponentially fast asymptotics of the De Sitter type but the energy density despite the unlimited expansion, is maintained as constant due to the dissipation process. For the closed model such evolutions never transform into the phase of contraction. The existence of the saddle S and of the knot K leads also to the appearance of specific solutions corresponding to the trajectories of the flat model SK and SO. The both solutions describe evolutions, free from cosmological singularities both in the beginning ($t=-\infty$) and in the end ($t=+\infty$). The solution was first described in [16]. However solutions of this type, as we see, correspond to separatrices of the saddles and are therefore unstable. And finally, note that the final stages of the evolution with unlimited expansion of the De Sitter type are possible now for the solutions caused by the creation of matter in the point I_1.

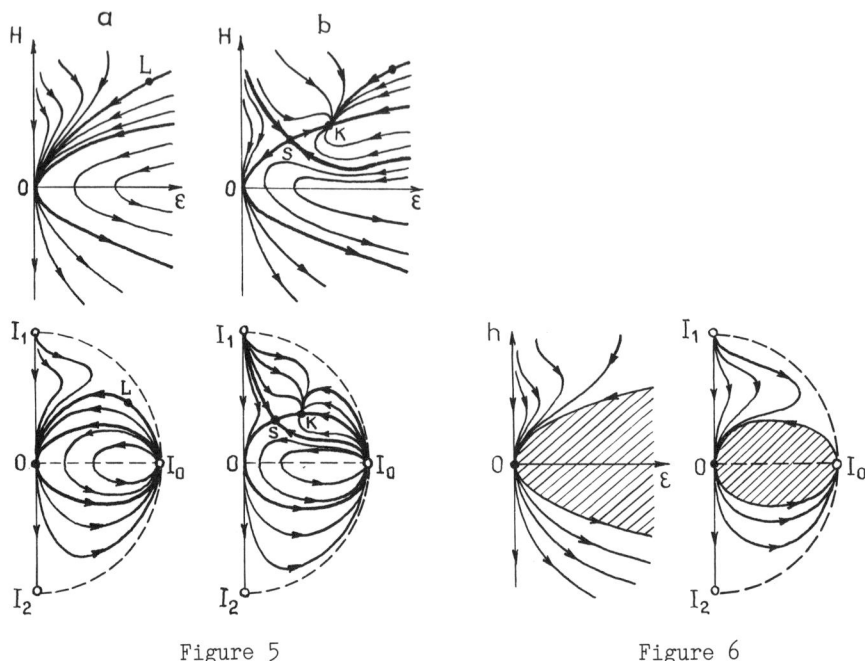

Figure 5   Figure 6

Now consider the main properties of the anisotropic cosmological model in the presence of dissipative effects due to viscosity. For this purpose let us take the homogeneous type-I model with the metric (1) and the energy-momentum tensor (44). Simple calculations testify that if alongside with the quantities r and h, determined by the relations (5), to introduce another ϕ from the condition

$$\eta = -\varphi/2, \tag{54}$$

then in the accompanying system the whole set of the Einstein equations (3) (again with $\Lambda=0$) and of the equations of hydrodynamics will read:

$$(\ln r_\alpha)^\cdot = h + s_\alpha r^{-3} e^\phi, \quad \sum_\alpha s_\alpha = 0 \tag{55}$$

$$\varepsilon = 3h^2 - l^2 r^{-6} e^{2\phi}, \tag{56}$$

$$\dot{h} = -3h(\varepsilon+p) + 9\zeta h^2 + 4\eta(3h^2-\varepsilon), \tag{57}$$

$$\dot{\varepsilon} = -3h(\varepsilon+p) + 9\zeta h^2 + 4\eta(3h^2-\varepsilon), \tag{58}$$

where s_a and l are the same constants as in (6)-(7). At the equation of state (10) and fixed dependences $\zeta(\varepsilon)$, $\eta(\varepsilon)$ Eqs.(57) and (58) again describe the dynamical system in the plane (ε,h):

$$\dot{\varepsilon} = -3\gamma\varepsilon h + 9\zeta h^2 + 4\eta(3h^2-\varepsilon),$$

$$\dot{h} = \frac{2-\gamma}{2}\varepsilon - 3h^2 + 3\zeta h/2. \tag{59}$$

Properties of this system were analysed in [7]. From the analysis of this work it ensues that the action of the bulk viscosity in anisotropic cosmology does not practically differ from what occurs in isotropic models. In this connection we shall describe only the results which are revelant to this aspect of the problem.

As it follows from the above mentioned estimates and considerations of a physical character it is natural to assume that the bulk viscosity coefficient ζ in the regions of small and large values of ε has the same asymptotics (50) and the shear viscosity coefficient η in these regions are approximated by power functions with the following requirements for the exponent:

$$\eta = \eta_1 \varepsilon^{\mu_1}, \quad \mu_1 \geq 1 \quad (\text{at } \varepsilon \to 0),$$

$$\eta = \eta_2 \varepsilon^{\mu_1}, \quad \mu_2 \geq \nu_2 + 1/2 \quad (\text{at } \varepsilon \to \infty). \tag{60}$$

(η_1 and η_2 are again arbitrary positive constants). Now we can analyse the behaviour of the trajectories of the system (59) in the vicinity of the boundaries of the phase plane $\varepsilon \to 0$ and $\varepsilon \to \infty$ and then consider, as previously, various versions of the matching of the obtained edges depending on the character of equilibrium states between these boundaries. Yet, it is clear that if these equilibrium points do exist, they are only due to the presence of bulk viscosity, and we shall assume that there are no equilibrium states between the regions $\varepsilon \to 0$ and $\varepsilon \to \infty$ whatsoever. Under these conditions the pattern of the trajectories of the system (59) will be as is given by Fig.6. Here two important circumstances are noteworthy. First, again the matter is created in the point I_1 and this effect is not sensitive to the fact by which type of viscosity it is generated. Yet, for instance, near $I_1 \eta >> \zeta$. Then at $\mu_1 > 1$ we get the following asymptotics of the solutions in this region ($t \to 0$ and $t > 0$):

$$r = \left(\sqrt{3}\, lt\right)^{1/3}, \quad h = 1/3t, \quad \varepsilon = \text{const} \cdot (3t/4\eta_1)^{1/(\mu_1-1)},$$

$$r_\alpha = \left(\sqrt{3}\, lt\right)^{P_\alpha},$$

$$p_\alpha = 1/3 + s_\alpha/\sqrt{3}\; 1,$$

$$p_1 + p_2 + p_3 = p_1^2 + p_2^2 + p_3^2 = 1 \qquad (61)$$

(the constant in this expression for ε is expressed in some way by the constant γ, μ_1). At $\mu_1 = 1$ for the quantities r, h and r_α the same expressions are still valid and for the energy density instead of (61) we get

$$\varepsilon = \text{const} \cdot (3t/4\eta_1)^{-\gamma} \exp(-4\eta_1/3t) \quad (\mu_1 = 1). \qquad (62)$$

The second noteworthy effect is a strong isotropizing action of the shear viscosity as the singular point I_o is being approached (contraction)>. Let this be a moment when t tends to zero from the side of the negative values of time. Then the asymptotics of the solution describing the contraction, near I_o (t→0 and t<0) will read:

$$r_\alpha = (-t)^{2/3\gamma}, \quad r = (-t)^{2/3\gamma}, \quad h = 2/3\gamma t, \quad \varepsilon = 4/3\gamma^2 t^2, \qquad (63)$$

i.e., purely Friedman. All the trajectories at t→0 asymptotically merge with the parabola $\varepsilon = 3h^2$ exponentially fast:

$$1 - \varepsilon/3h^2 \sim \exp\left[\rho^2 t(-t)^{-2\mu_2}\right]. \qquad (64)$$

This strong isotropization in the vicinity of the singularity is associated with the fact that in the upper half plane (h>0) there is no trajectory, except Friedman, which would correspond to the solution with the Big Bang and to an infinite value of the energy density in the beginning of the evolution. All trajectories now start with the vacuum Kasner singularity in the point I with the zero value in the beginning of the evolution.

ACKNOWLEDGEMENT

The author is thankful to Dr.V.Belinsky for his assistance in the preparation of the manuscript. Part of the results has been obtained in cooperation with him.

DEDICATION

The author dedicates this article to Professor Peter Bergmann, one of the founders of the contemporary relativity theory.

REFERENCES

1. I. S. Shikin, The homogeneous axis-symmetric model in the ultra-relativistic case, Doklady Akademii Nauk **176**:1048-1051 (1967).
2. C. B. Collins, More qualitative cosmology, Comm. Math. Phys. **23**:137-144 (1971).
3. V. A. Belinsky, E.M.Lifshitz, I.M.Khalatnikov, Oscillatory regime of the approach to the singular point in relativistic cosmology, Uspekhi Fiz. Nauk **102**:463-500 (1970).
4. V. A. Belinsky, E. M. Lifshitz, I. M. Khalatnikov, On the construction of the general cosmological solution of the Einstein equation with the singularity in time, ZhETF **62**:1606-1613 (1972).

5. O. I. Bogoyavlensky, S. P. Novikov, Peculiarities of the cosmological Biancchi-IX model in terms of the qualitative theory of differential equations, ZhETF **64**:1475-1494 (1973).
6. O. I. Bogoyavlensky, "Methods of the qualitative theory of dynamical systems in astrophysics and gas dynamics", Nauka, Moscow (1980).
7. V. A. Belinsky, I.M.Khalatnikov, On the influence of viscosity on the character of cosmological evolution, ZhETF **69**:401-413 (
8. V. A. Belinsky, I. M. Khalatnikov, Viscosity effects in isotropic cosmology, ZhETF **72**:3-17 (1977).
9. E. S. Nikomarov, I. M. Khalatnikov, Qualitative isotropic cosmology with the cosmological constant with the dissipation taken into account, ZhETF **75**:1176-1180 (1978).
10. V. A. Belinsky, E. S. Nikomarov, I. M. Khalatnikov, The study of cosmological evolution of the viscous-elastic matter with causal thermodynamics ZhETF N 2:417-433 (1979).
11. A. A. Andronov, E. A. Leontovich, I. I. Gordon, A. G. Mayer, "The theory of bifurcations of dynamical systems on the plane", Nauka, Moscow (1967).
12. A. A. Andronov, E. A. Leontovich, I. I. Gordon, A. G. Mayer, "Qualitative theory of the second-order dynamical systems", Nauka, Moscow (1966).
13. V. A. Belinsky, I. M. Khalatnikov, On the influence of scalar and vector fields on the character of the cosmological singularity, ZhETF **63**:1121-1134 (1972).
14. V. A. Belinsky, I. M. Khalatnikov, On the influence of matter and physical fields upon the nature of cosmological singularities, in: "Soviet Physics Reviews", Harwood Ac. Publ., Switzerland (1981).
15. L. D. Landau, E. M. Lifshitz, "Mechanics of continuous media", Fizmatgiz, Moscow (1953).
16. G. L. Murphy, Big Bang model without singularities, Phys. Rev. D, **8**:4231-4233 (1973).

THIRD QUANTIZATION OF GRAVITY AND THE COSMOLOGICAL CONSTANT PROBLEM

G. Lavrelashvili
Tbilisi Mathematical Institute, 380093 Tbilisi, USSR
and
V.A. Rubakov and P.G. Tinyakov
Institute for Nuclear Research of the Academy of Sciences of the USSR
60th October Anniversary Prospect 7a Moscow, 117312, USSR

1 Introduction

The cosmological constant problem is one of the most intriguing problems of modern physics (for a review see, e.g., ref.[1]). A proposal for its solution due to Baum [2] and Hawking [3], and its modification by Coleman [4], have attracted much attention. The arguments of refs.[2-4] have essentially two crucial ingredients:

i) One assumes that the observable value of the cosmological constant, Λ, is not an absolutely fundamental c-number parameter of the theory; rather, one has to consider universes with different values of Λ which are distributed according to some distribution function $P(\Lambda)$. In this sense Λ is a dynamical (quantum) variable. This assumption has been supported by the recent study of topological changes (baby universes, wormholes) [5-7] which has finally led to the observation [8-10] that all coupling constants, including Λ, are dynamical variables in the above sense.

ii) It is assumed that the probability distribution, $P(\Lambda)$, is calculable. Furthermore, it is assumed that $P(\Lambda)$ is saturated by the contribution of the De Sitter instanton (4-sphere) into the euclidean functional integral in quantum gravity. This instanton exists at $\Lambda > 0$ and has negative action, $I_{inst} = -3\pi M_{Pl}/\Lambda$. Thus, Baum [2] and Hawking [3] argued that

$$P(\Lambda) \propto \exp(-I_{inst}) = \exp(3\pi M_{Pl}/\Lambda) , \qquad (1.1)$$

i.e., $P(\Lambda)$ is strongly peaked at $\Lambda = 0$, as desired. Coleman [4] proposed to sum up the multi-instanton contributions and obtained even stronger peak,

$$P(\Lambda) \propto e^{\exp(-I_{inst})} = e^{\exp(3\pi M_{Pl}/\Lambda)} \qquad (1.2)$$

Although these arguments have several attractive features, they appear too formal. Moreover, they say nothing about the probability distribution at *negative* Λ, and one may wonder whether negative values of Λ are even more probable than zero.

It is certainly of importance to have the interpretation of the results like eqs.(1.1) or (1.2), which would not be based entirely on the euclidean functional integral of gravity.

One possible interpretation [11,12] is that $P(\Lambda)$ is the number of large universes created in superspace from the state containing no small (Planck size) universes. This suggestion is supported by an analogous interpretation of the negative euclidean action instantons in ordinary relativistic quantum mechanics/quantum field theory [13]. Another interpretation, coinciding technically but not conceptually with that of refs.[11,12], has been suggested in refs.[14,15] where $P(\Lambda)$ has been associated with the magnitude of the wave function of the universe obeying certain boundary conditions at small (Planck) sizes.

An advantage of both these approaches is that they allow one, at least in principle, to address the problem of the most probable matter content in the universe, as well as the question of the negative cosmological constant.

The purpose of this paper is to review the third quantization approach and discuss these two problems in some detail. This paper is organized as follows. We consider in sect.2 the case $\Lambda > 0$ and review the arguments of refs.[11,12]. To recall the concept of creation of universes in (mini)superspace [16,11] and introduce notations, we consider in sect.2.1 the simplest minisuperspace model where the only variable is the radius of the universe. We introduce matter variables in sect.2.2. We do not assume the matter fields to be homogeneous and isotropic; instead, we treat them as perturbations and use the approximation where their back reaction on the dynamics of the radius of the universe is neglected. This is the principal approximation to be used in this paper. While in the case of positive Λ this approximation works well, at negative Λ it is not entirely self-consistent. However, we think that our arguments are sufficiently general and are not merely artifacts of the approximation, which, at any rate, is expected to be qualitatively correct.

In sect.2.2 we evaluate the matter content of the universes created in (mini)superspace at small positive Λ. To be specific, we study massive scalar particles conformally coupled to gravity. We find that the number density of particles and their energy density are small, so that the created universes are mostly empty.

We turn to the problem of negative Λ in sect.3. We estimate the number of classical universes created in minisuperspace from the state that contains no small (Planck size) universes. This number determines the distribution function $P(\Lambda)$. In fact, we consider the number of created classical universes at given value of the radius, a_0, which corresponds to the conditional probability $P(\Lambda, a_0)$ to observe the cosmological constant Λ provided that the universe has the size a_0. We find that $P(\Lambda, a_0)$ is not peaked at small $|\Lambda|$; instead, it exponentially increases as $(-\Lambda)$ gets large. So, we conclude that large negative Λ is more probable than zero. Clearly, this observation creates a problem for Baum-Hawking-Coleman mechanism. We point out in sect.4 a possible solution to this problem that invokes the weak anthropic principle.

2 Creation of universes in minisuperspace

2.1 The simplest minisuperspace model

In this section we consider creation of universes in the simplest minisuperspace model with just one dynamical variable, namely, the radius, a, of the closed Friedmann-Robertson-Walker universe.

We start from the Wheeler-DeWitt equation for the wave function of the universe which, in this model, is a function of a only. This equation has the following form,

$$\left(\frac{1}{2}\frac{\partial^2}{\partial a^2} + \frac{1}{2}\mu^2(a)\right)\Psi(a) = 0, \tag{2.1}$$

where
$$\mu^2(a) = -a + \frac{1}{3}\Lambda a^3 + \varepsilon,$$

Λ is the cosmological constant and ε is some positive number introduced for convenience (in the end of the calculations we take the limit $\varepsilon \to 0$). Hereafter we use the units in which $3\pi M_{Pl}^2/2 = 1$. In this section we assume $0 < \Lambda \ll 1$.

We tend to use the analogy between eq.(2.1) and the Klein-Gordon equation in (0+1) dimensions, a being the analog of time and $\mu^2(a)$ being the time dependent (mass)^2. μ^2 is positive in two regions : i) small a ($0 < a < a_1$) and ii) large a ($a \geq a_1$), where $a_1 \simeq \varepsilon^{1/2}$, $a_2 \simeq \Lambda^{-1/2}$ are the two turning points. Accordingly, we can define two oscillating semiclassical solutions in each of these regions,

$$\Psi_{in}^{\pm}(a) = (2\mu)^{-1/2}\exp(\pm S_{in}), \qquad a < a_1$$

$$\Psi_{out}^{\pm}(a) = (2\mu)^{-1/2}\exp(\pm S_{out}), \qquad a > a_2,$$

where
$$S_{in} = \int_0^a \mu(a')da',$$

$$S_{out} = \int_{a_2}^a \mu(a')da'.$$

The standard interpretation [17-21] of Ψ_{in} and Ψ_{out} is that they describe classical Friedmann-like and De Sitter-like universes, respectively.

In the intermediate region $a_1 < a < a_2$ the two linear independent semiclassical solutions to eq.(2.1) exponentially increase and decrease,

$$\Psi_{incr}(a) = (2|\mu|)^{-1/2}e^{I(a)}$$

$$\Psi_{decr}(a) = (2|\mu|)^{-1/2}e^{-I(a)} \tag{2.2}$$

where
$$I(a) = \int_{a_1}^a |\mu(a')|da' \tag{2.3}$$

The solutions (2.2) have no interpretation in terms of classical universes; the absence of oscillating solutions corresponds to the fact that no homogeneous and isotropic solutions to the Einstein equations exist in the region $a_1 < a < a_2$.

To obtain the minisuperspace version of the third quantized gravity [22-24,16,11,12], one considers $\Psi(a)$ as an operator. To distinguish between the c-number and operator wave functions of the universe, we denote latter by $\hat{\Psi}(a)$. The equal "time" commutation relations consistent with eq.(2.1) are

$$[\hat{\Psi}^{\dagger}, \partial_a\hat{\Psi}] = i,$$

$$[\hat{\Psi}^{\dagger}, \hat{\Psi}] = [\partial_a\hat{\Psi}^{\dagger}, \partial_a\hat{\Psi}] = 0.$$

At the moment it is not clear whether $\hat{\Psi}$ is hermitean or not; we assume for definiteness that $\hat{\Psi}$ is hermitean, although this assumption is not essential for our purposes. Then in the regions $a < a_1$ and $a > a_2$, the operator $\hat{\Psi}(a)$ can be represented as follows,

$$\hat{\Psi} = \hat{A}\Psi_{in}^{(-)}(a) + \hat{A}^{\dagger}\Psi_{in}^{(+)}(a) \qquad a < a_1,$$

$$\hat{\Psi} = \hat{B}\Psi_{out}^{(-)}(a) + \hat{B}^{\dagger}\Psi_{in}^{(+)}(a) \qquad a > a_2,$$

where the a-independent operators $\hat{A}$, $\hat{A}^\dagger$ and $\hat{B}$, $\hat{B}^\dagger$ obey the standard commutation relations

$$[\hat{A}, \hat{A}^\dagger] = [\hat{B}, \hat{B}^\dagger] = 1 \ .$$

The operator $\hat{A}^\dagger$ is interpreted as the creation operator of small Friedmann-like universes, while $\hat{B}^\dagger$ creates large De Sitter-like ones.

The state without small universes is the $\hat{A}$-vacuum $\|0\rangle\rangle_F$,

$$\hat{A}\|0\rangle\rangle_F = 0 \ .$$

This state, however, contains certain number of De Sitter-like universes,

$$N_{DS} = {}_F\langle\langle 0\|\hat{B}^\dagger \hat{B}\|0\rangle\rangle_F \ .$$

N_{DS} can be called as the number of De Sitter-like universes created in minisuperspace from the state containing no Friedmann-like ones, and the calculation of N_{DS} coincides technically with the calculation of pair creation in (0+1)-dimensional scalar field theory in time-dependent background field $\mu^2(a)$. The operators $\hat{A}$ and $\hat{B}$ are related by the Bogoliubov transformation

$$\hat{B} = u\hat{A} + v\hat{A} \ ,$$

where

$$v = i[\Psi_{out}^{(-)}]^* \overleftrightarrow{\partial}_a \Psi_{in}^{(+)} \ .$$

The number of created universes is

$$N_{DS} = |v|^2 \ .$$

To perform the actual calculation of v, one has to continue $\Psi_{in}^{(-)}$, via eq.(2.1), to the region $a > a_2$ and then take its projection onto $\Psi_{out}^{(+)}$. In the region $a_1 < a < a_2$, the function $\Psi_{in}^{(-)}$ contains both exponentially increasing and decreasing parts,

$$\Psi_{in}^{(-)} = \frac{1}{2}\Psi_{incr} + \frac{1}{2}\Psi_{decr} \tag{2.4}$$

At small Λ, only the increasing part survives at $a \sim a_2$. At $a > a_2$, the function Ψ_{incr} contains $\Psi_{out}^{(+)}$ and $\Psi_{out}^{(-)}$ with equal weights, while its magnitude is proportional to

$$\Psi_{incr}(a_2) \propto e^{-1/2 I_{inst}} \ ,$$

where

$$\frac{1}{2} I_{inst} = -\int_{a_1}^{a_2} |\mu| da \ .$$

Therefore, $v \propto \Psi_{incr}(a_2)$ and, finally,

$$N_{DS} = e^{-I_{inst}} \ .$$

The notation I_{inst} is not accidental: I_{inst} is equal to the euclidean action of the theory evaluated at the periodic solution to the euclidean Einstein equations (instanton). In the limit $\varepsilon \to 0$, the explicit formulae for $I(a)$ and I_{inst} are

$$I(a) = -\frac{1}{\Lambda}\left((1 - \frac{\Lambda}{3}a^2)^{3/2} - 1\right) \tag{2.5}$$

and

$$I_{inst} = -\frac{2}{\Lambda} \qquad (2.6)$$

Note that in this limit the relevant solution is the De Sitter instanton and I_{inst} is its action. Thus, the number of created universes is

$$N_{DS} = e^{2/\Lambda} \qquad (2.7)$$

which is equal to the Baum-Hawking exponential factor (recall, that we use the units $3\pi M_{\text{Pl}}^2/2 = 1$).

Now let us suppose that Λ is a "dynamical variable", i.e. that one has to consider the ensemble of theories with different cosmological constants. Let the amplitude to have a theory with the cosmological constant Λ be $f(\Lambda)$. Let us also assume that the state of the system of universes (the state of the googolplexus [12]) is the Λ-vacuum containing no small universes. This choice may be the most natural one, especially in view of the idea of the creation of the universes from nothing [25,26]. The quantity we wish to calculate is the conditional probability $P(\Lambda, a_0)$ to pick up a classical universe with the cosmological constant Λ under the condition that the radius of the universe is equal to a_0 [15].

In the simplest minisuperspace model of this section, the calculation is very simple. At $\Lambda < 3/a_0^2$, no classical universes of the radius a_0 exist, so $P(\Lambda, a_0) = 0$. At $\Lambda > 3/a_0^2$ the probability to pick up a universe with given Λ is proportional to the number of such universes, N_{DS}. So

$$P(\Lambda, a_0) = \text{const}|f(\Lambda)|^2 e^{2/\Lambda}, \qquad \Lambda > 3/a_0^2,$$

$$P(\Lambda, a_0) = 0, \qquad \Lambda < 3/a_0^2.$$

At large a_0, the probability has a sharp maximum at

$$\Lambda = \Lambda_{max} = 3/a_0^2,$$

and its maximum value is

$$P(\Lambda, a_0) = \text{const}|f(\Lambda)|^2 e^{\frac{2}{3}a_0^2}. \qquad (2.8)$$

This conclusion coincides with that of ref.[15] and is valid, of course, if $f(\Lambda)$ is a smooth non-vanishing function near $\Lambda = 0$ (cf. ref.[4]).

If one drops the requirement that the radius of the universe is fixed, then the probability to pick up a large universe is determined solely by N_{DS},

$$P(\Lambda) = \text{const}|f(\Lambda)|^2 e^{2/\Lambda}.$$

This is precisely the Baum-Hawking [2,3] expression.

2.2 Turning on matter degrees of freedom

In this section we extend the simplest model of sect.2.1 by introducing matter degrees of freedom. There are two technically tractable ways to include matter into the minisuperspace models: i) one takes into account the homogeneous and isotropic degrees of freedom only and obtains more complicated Wheeler-DeWitt equation which, however, still contains finite number of degrees of freedom (see, e.g., refs.[27-29]); ii) one treats the matter Hamiltonian as the perturbation and arrives at the problem of the behaviour of the infinite number of matter degrees of freedom (and/or gravitons) on a given homogeneous and isotropic gravitational background [30-32,20]. In this paper we follow the second route with slight modification which is needed because we are mainly interested in the region $a_1 < a < a_2$ where no classical background universes exist (cf.[20,33]).

The Wheeler-DeWitt equation for the minisuperspace model with matter reads

$$\left(\frac{1}{2}\frac{\partial^2}{\partial a^2} + \frac{1}{2}\mu^2(a) + aH_M\right)\Psi(a) = 0 ,\tag{2.9}$$

where H_M is the matter Hamiltonian defined with respect to proper time. Specifically, we consider massive scalar field conformally coupled to gravity, whose lagrangian is

$$\mathcal{L} = -(g)^{1/2}\left(\frac{1}{2}(\partial_\mu\varphi)(\partial^\mu\varphi) - \frac{\mathcal{R}}{12}\varphi^2 - \frac{m^2}{2}\varphi^2\right) .\tag{2.10}$$

It is convenient to use the conformally rescaled field $\phi = a\varphi$ and its conjugate momentum π_ϕ. It is also convenient to decompose ϕ and π_ϕ over the spherical harmonics on a unit 3-sphere, ϕ_{nlm} and π_{nlm}, $n = 1, 2, ...; l = 0, ..., (n-1); m = -l, ..., l$. Then the matter Hamiltonian defined with respect to the *conformal* time is

$$\mathcal{H} = aH_M = \sum_{n=1}^{\infty}\sum_{l,m}\left(\frac{1}{2}\pi_{nlm}^2 + \frac{1}{2}\omega_n^2\phi_{nlm}^2\right) ,\tag{2.11}$$

where

$$\omega_n^2 = n^2 + m^2 a^2 .\tag{2.12}$$

One possible representation of the standard commutation relations between ϕ_{nlm} and π_{nlm} is the "coordinate" representation. In this representation, the wave function of the universe Ψ is a function of ϕ_{nlm} and a. We will again use the analogy between Wheeler-DeWitt equation equation and the Klein-Gordon equation, the coordinates of the minisuperspace, a and ϕ_{nlm} being the analogs of time and space coordinates, respectively, while

$$M^2(a, \phi_{nlm}) = \mu^2(a) + \sum \omega_n^2(a)\phi_{nlm}^2 \tag{2.13}$$

being the analog of the space-time dependent (mass)^2.

In the third quantized gravity, the wave function $\hat{\Psi}(a, \phi_{nlm})$ becomes an operator with the standard equal "time" commutation relations. At small a, it can be expressed in terms of the operators $\hat{A}^\dagger_{\{k\}}$ and $\hat{A}_{\{k\}}$ that create and annihilate universes with the matter content $\{k\}$. Namely, the eigenfunctions of the matter conformal Hamiltonian are the wave functions of the state of the scalar field that contains k_{nlm} particles in modes with quantum numbers (n, l, m), respectively. Their conformal energies are

$$\mathcal{E}_{\{k\}} = \sum_{nlm}(\frac{1}{2} + k_{nlm})\omega_n(a) .\tag{2.14}$$

In general, the corresponding states depend also on a (because ω_n^2 depends on a), but at small a this dependence disappears.

The decomposition of the operator $\hat{\Psi}$ at small a has the following form

$$\hat{\Psi}(a, \phi) = \sum_{\{k\}}\hat{A}^\dagger_{\{k\}}e^{i\Omega_{\{k\}}a}\psi_{\{k\}}(\phi_{nlm}) + \text{h.c.} ,\tag{2.15}$$

where

$$\Omega^2_{\{k\}} = \varepsilon + 2\mathcal{E}_{\{k\}} .$$

The Friedmann vacuum $\|0\rangle\rangle_F$ is again defined by the requirement that $\hat{A}_{\{k\}}\|0\rangle\rangle_F = 0$ for all k. It contains no small universes.

The decomposition analogous to eq.(2.15) can be written also at large a, the corresponding creation and annihilation operators being $\hat{B}^\dagger_{\{k\}}$ and $\hat{B}_{\{k\}}$, respectively. We can again consider the problem of the creation of universes in (mini)superspace.

The creation of universes occurs most efficiently in the region of minisuperspace where the effective (mass)^2 is negative,

$$M^2(a, \{\phi\}) < 0 \,.$$

In this region, the system is unstable, and the creation of universes is exponential. Clearly, the instability is strongest near $\{\phi\} = 0$, so we can use the semiclassical techniques for evaluating the Bogoliubov coefficients, as outlined below.

The leading exponential factor in the number of created universes is determined by the exponentially increasing solution to eq.(2.9) in the region $a_1 \lesssim a \lesssim a_2$. We consider the matter hamiltonian as the perturbation and write the solution in the following form,

$$\Psi = e^{I(a)} \tilde{\Psi}(a, \{\phi\}) \,, \tag{2.16}$$

where $I(a)$ is the unperturbed euclidean action, eq.(2.3). Inserting eq.(2.16) into eq.(2.9) and neglecting the terms of the higher orders in H_M, we obtain the euclidean Schrödinger equation, (cf.[20,30-32])

$$\frac{\delta \tilde{\Psi}}{\delta \eta} = -(a H_M) \tilde{\Psi} \,, \tag{2.17}$$

where the new variable η is related to a via

$$\frac{\delta a}{\delta \eta} = \frac{\delta I}{\delta a} \,. \tag{2.18}$$

η can be called the euclidean conformal time, because eq.(2.18) is equivalent to the unperturbed ($H_M = 0$) euclidean Einstein equation for the following choice of the metrics

$$ds^2 = a^2(\eta)(d\eta^2 + d\Omega_3^2) \,.$$

Recall, that $\mathcal{H} \equiv aH_M$ is the matter conformal Hamiltonian. Within our approximation, the problem of the creation of universes from the Friedmann vacuum $\|0\rangle\rangle_F$ reduces to that of solving eq.(2.17) with the initial condition

$$\tilde{\Psi}(a, \{\phi_{nlm}\})\Big|_{a=a_1} = \psi_{\{k\}}(\{\phi_{nlm}\}) \,.$$

The Bogoliubov coefficients $v_{\{k\},\{q\}}$ are determined by the projection of $\tilde{\Psi}(a=a_2)$ onto the eigenfunctions of the conformal Hamiltonian at $a = a_2$.

The number of created universes with the matter content $\{q\}$ is

$$N_{\{q\}} = \sum_{\{k\}} |v_{\{k\},\{q\}}|^2 \tag{2.19}$$

To solve eq.(2.17), we first note that the wave function $\tilde{\Psi}(a, \{\phi_{nlm}\})$ factorizes,

$$\tilde{\Psi}(a, \{\phi_{nlm}\}) = \prod_{nlm} \tilde{\Psi}_{nlm}(a, \phi_{nlm}) \,.$$

Therefore, we can solve eq.(2.17) mode by mode. In what follows we omit the subscript (n, l, m) and consider the single mode euclidean Schrodinger equation. Second, we use the Fock representation for the wave function of each mode, instead of the coordinate representation. Namely, at every time η, one can define the conformal Hamiltonian for each mode whose eigenfunctions, $\psi_k(\phi|\eta)$, can be interpreted as the states with k particles in the mode under study. The dependence of ψ_k on η originates from the time dependence of ω_n^2. This interpretation is valid provided that $\dot{\omega}_n/\omega_n$ is small, which is the case at small Λ. We introduce the q-particle amplitudes

$$C_q(\eta) = (\psi_q(\eta), \check{\Psi}(\eta)) \ .$$

It follows from eq.(2.17) that C_q obey

$$\dot{C}_q = -\omega_n(q + \frac{1}{2})C_q + \frac{\dot{\omega}_n}{4\omega_n}\sqrt{q(q-1)}C_{q-2} - \frac{\dot{\omega}_n}{4\omega_n}\sqrt{(q+1)(q+2)}C_{q+2} \qquad (2.20)$$

To obtain the Bogoliubov coefficients, we have to solve eq.(2.20) with the initial condition

$$C_q^{(k)}\Big|_{a=a_1} = \delta_{k,q} \ ,$$

where k is the initial number of particles in the mode (n, l, m) in the virtual Friedmann universe. The Bogoliubov coefficients $v_{\{k\},\{q\}}$ are then

$$v_{\{k\},\{q\}} = e^{I(a)} \prod_{nlm} C_q^{(k)}(a_2) \ .$$

The first qualitative feature of eq.(2.20) is that $C_q^{(k)}(a_2)$ is maximal at $k = 0$ (no particles in the virtual Friedmann universe). Therefore, we concentrate on the case $k = 0$ and omit the superscript k in what follows.

At $k = 0$, we can solve eq.(2.20) perturbatively, the small parameter being $\dot{\omega}/\omega$. To the zeroth order in this parameter we have

$$C_0(a) = \exp\left(-\frac{1}{2}\int_{\eta(a_1)}^{\eta(a)} \omega_n(\eta)d\eta\right) \ ,$$

$$C_q(a) = 0 \ , \qquad q > 0 \ .$$

The non-trivial behaviour of $C_0(a)$ is due to the zero-point (conformal) energy of the scalar field and is irrelevant for our purposes. To the first non-trivial order in $\dot{\omega}/\omega$ we obtain the amplitude of the creation of a pair of scalar particles in a given mode

$$\frac{C_2(a)}{C_0(a)} = \int_{\eta(a_1)}^{\eta(a)} \exp\left\{-2\int_{\eta'}^{\eta(a)} \omega_n d\eta''\right\} \frac{\sqrt{2}}{4}\frac{\dot{\omega}}{\omega}(\eta')d\eta \qquad (2.21)$$

The quantity of interest is the amplitude to have a pair in the De Sitter-like universe, $(C_2/C_0)(a_2)$. At $a = a_2 = (3/\Lambda)^{1/2}$ the integral in eq.(2.21) is saturated near the upper limit, and its value is

$$\frac{C_2(a)}{C_0(a)} = \frac{\sqrt{2}}{4}\frac{(ma_2)^2}{(n^2 + m^2 a_2^2)^2} \qquad (2.22)$$

This is a small number, provided that

$$m^2 a_2^2 \equiv \frac{3m^2}{\Lambda} \gg 1$$

Making use of eq.(2.22) we can calculate the average number of pairs that exist in the universe created in superspace,

$$\nu_{pairs} = \sum_{nlm} |\frac{C_2(a)}{C_0(a)}|^2$$

We find

$$\nu_{pairs} \sim \frac{1}{ma_2} \ll 1$$

Thus, we conclude that the large universes created in the superspace are mostly empty.

3 The problem of negative Λ

In this section we consider the number, $N(\Lambda, a_0)$, of classical universes of the size a_0 created in a theory with the negative cosmological constant Λ. As before, we assume that the state of the system of universes (the googolplexus) is such that there are no small (Planck size) universes, i.e., that this state is the Friedmann vacuum, $\|0\rangle\rangle_F$. If the amplitude to have a theory with the cosmological constant Λ is $f(\Lambda)$, then the distribution function (introduced in sect.2) to pick up a universe of the size a_0 with a given value of the cosmological constant is

$$P(\Lambda, a_0) = |f(\Lambda)|^2 N(\Lambda, a_0) \ .$$

If, as before, $f(\Lambda)$ is a smooth non-vanishing function, then the distribution $P(\Lambda, a_0)$ is determined essentially by $N(\Lambda, a_0)$.

In the simplest model of sect.2.1, the number of created classical universes is zero at negative Λ: there are no large classical universes in this model. However, $N(\Lambda, a_0)$ is nonzero at large a in any realistic theory containing matter degrees of freedom and/or gravitons. In that case the classical evolution of the universe is possible provided that the energy of matter is large enough (hereafter we consider the case $\Lambda a_0^2 \gg 1$), i.e. the conformal energy of matter is larger than the conformal energy associated with the cosmological constant,

$$\mathcal{E} > \frac{1}{6}|\Lambda|a_0^4 \qquad (3.1)$$

To be specific, we again consider the model of matter determined by the lagrangian (2.10). At a near a_0, there exist two types of solutions to the Wheeler-DeWitt equation (2.9): i) the oscillating ones and ii) the exponentially increasing or decreasing ones. The latter solutions do not describe any classical universes. The former ones correspond to the classical universes; the matter content of these universes obeys, of course, eq.(3.1), where the conformal energy of matter is given by eq.(2.14). As before, one can introduce the creation and annihilation operators associated with the oscillating solutions; they are related to the operators creating and annihilating the Friedmann-like universes via the Bogoliubov coefficients. These Bogoliubov coefficients determine the number of classical universes created at $a = a_0$ from the Friedmann vacuum. Repeating the arguments of sect.2.2 we find that in the approximation where the back reaction of matter on the dynamics of the scale factor is neglected, the Bogoliubov coefficients are

$$v_{\{k\},\{q\}} = e^{I(a_0)}(\tilde{\Psi}_{\{k\}}(a_0), \psi_{\{q\}}) \ ,$$

where $I(a)$ is given by eq.(2.5), $\tilde{\Psi}_{\{k\}}(a_0)$ is the solution to the euclidean Schrodinger equation and $\psi_{\{q\}}$ is the state vector of matter in the created classical universe. We recall again that the matter content $\{q\}$ should obey eq.(3.1).

The total number of created classical universes is, within our approximation,

$$N(a_0) = \sum_{\{k\},\{q\}} |v_{\{k\},\{q\}}|^2 .$$

Again, it is straightforward to see that the sum over $\{k\}$ is saturated by the term corresponding to the zero number of scalar bosons in the initial virtual universe. Now, it is natural to introduce the probability to create, during the euclidean evolution, scalar particles in such a way that thier total conformal energy exceeds $\mathcal{E}$,

$$p(a_0, \mathcal{E}) = \sum_{\substack{\{q\} \\ \mathcal{E}_{\{q\}} > \mathcal{E}}} |(\tilde{\Psi}_{\{0\}}(a_0), \psi_{\{q\}})|^2 \qquad (3.2)$$

In terms of this quantity, the number of created classical universes is

$$N(a_0) = e^{2I(a_0)} p(a_0, \frac{1}{6}\Lambda a_0^4) .$$

In what follows we estimate two contributions into the sum in eq.(3.2). The first contribution comes from the soft modes, $n \lesssim ma_0$. The conformal energy of a particle in each of these modes is $\omega \sim ma_0$ while the total number of these modes is $\Delta \sim (ma_0)^3$. For the conformal energy of matter be larger than or equal to $\mathcal{E}$, the number of scalar pairs created in these modes must be of order $\mathcal{E}/(ma_0)$. The probability to create a pair in one of these modes is of order (cf. eq.(2.22))

$$\delta = \left|\frac{C_2}{C_0}\right|^2 \sim \frac{|\Lambda|}{m^2} ,$$

(recall that we assume $|\Lambda|a_0^2 \gg 1$). We consider the case of small $|\Lambda|$, $|\Lambda|/m^4 \ll 1$, and also take m to be small (in Planck units), $m^2 \ll 1$. Then the probability to excite $\mathcal{E}/(ma_0)$ modes out of Δ can be estimated by

$$p(a_0, \mathcal{E}) = \exp\left(-\text{const}\frac{\mathcal{E}}{ma_0}[\ln(\frac{\mathcal{E}}{ma_0 \delta \Delta} + O(1)]\right) ,$$

where the constant is of order one (we will omit it in what follows).
If this contribution were the only one, then $p(a_0, \frac{1}{6}|\Lambda|a_0^4)$ would be roughly

$$p(a_0, \frac{1}{6}|\Lambda|a_0^4) = \exp\left(a_0^3 - \frac{|\Lambda|a_0^3}{6m} \ln \frac{1}{m^2}\right)$$

and the number of created classical universes would be

$$N(a_0, \Lambda) = \exp\left(\left|\frac{\Lambda}{27}\right|^{1/2} a_0^3 - \frac{|\Lambda|a_0^3}{6m} \ln \frac{1}{m^2}\right)$$

This expression is valid in the region $|\Lambda| \ll m^4$. It is straightforward to see that $N(a_0, \Lambda)$ monotonically increases in this region and

$$N(a_0, |\Lambda| \sim m^4) = \exp\left((m^2 - m^3 \ln \frac{1}{m^2})a_0^3\right)$$

where we omit all factors of order one in the exponential. This value is to be compared with eq.(2.8). We see that the number of created universes is much larger at negative

Λ than at positive Λ, provided that $ma \gg 1$. Thus, we see the effect of the soft modes is sufficient to make the cosmological constant probably negative.

Another contribution into the sum in eq.(3.2) which is even more important at large negative Λ, comes from the hard modes, $n \sim \mathcal{E}$. Namely, let us consider the probability of creation of k pairs of particles with conformal energies of order $\mathcal{E}/k$. The probability to create one such pair is of order (see eq.(2.22))

$$\left|\frac{C_2(a)}{C_0(a)}\right|^2 \sim \frac{|\Lambda|m^4 a_0^6}{n^6} ,$$

where $n \sim \mathcal{E}/k$. The phase space is estimated by

$$\sum_{nlm} \sim n^3 .$$

Thus, the contribution under discussion is roughly equal to

$$p(a_0, \mathcal{E}) \sim \left(\frac{|\Lambda|m^4 a_0^6}{n^3}\right)^k$$

At $\mathcal{E} \sim |\Lambda| a_0^4$ we have

$$p(a_0, \frac{1}{6}|\Lambda|a_0^4) \sim \left(\frac{m^4 k^3}{|\Lambda|^2 a_0^6}\right)^k \tag{3.3}$$

and the number of created classical universes is

$$N(a_0, \Lambda) = \exp\left(\left|\frac{\Lambda}{27}\right|^{1/2} a_0^3 - \frac{|\Lambda|a_0^3}{6m} k \ln \frac{\Lambda^2 a_0^6}{m^4 k^3}\right) \tag{3.4}$$

As is clear from eq.(3.3), at fixed k of order one, the creation of matter sufficient for the universe to materialize is only power suppressed at large $|\Lambda|$, while the original growth of the wave function, eq.(2.2), is exponential. Accordingly, the number of created universes, eq.(3.4), has a strong peak at $\Lambda = -\infty$. Thus, the interpretation of the Baum-Hawking result in terms of the creation of universes in superspace is likely to imply that the cosmological constant is very probably negative and infinite.

4 Conclusion

The universes predominantly created in superspace either at small positive Λ or at large negative Λ are very different from the universe we live in. At $0 < \Lambda \ll 1$ the created universes are empty, while at $\Lambda \to -\infty$ they contain a few very energetic particles. Neither of these types of universes is suitable for the existence of intelligent observers. Therefore, the quantity $N(a_0, \Lambda)$ we attempted to calculate in this paper is not, in fact, directly relevant for the discussion of the consistency of the theory with observations. The quantity of interest is the number of universes (created in superspace) where life of our type is possible. These anthropic considerations, which seem very natural in the third quantized gravity, may help significantly to avoid the unpleasant conclusions seemingly following from the results of this paper.

At $0 < \Lambda \ll 1$, a possible scenario [11] consistent with both the anthropic principle and the Baum-Hawking exponential behaviour is that the most probable anthropogenic universes created in superspace are large ($a \sim 1/\Lambda^{1/2}$) almost empty ones, containing homogeneous scalar field in a finite region of space. At small Λ, this region occupies a negligible portion of the whole universe, so the exponential behaviour (2.8) is likely to remain unaltered. The homogeneous scalar field can drive the chaotic inflation [34] that eventually lead to the occurrence of life.

As to the large negative values of the cosmological constant, they have been argued to be inconsistent with the anthropic principle [1]. Furthermore, the anthropic bound at $\Lambda < 0$ is close to the observational limit [1]. Whether the arguments of this paper together with ref.[1] provide the solution to the cosmological constant problem is yet to be understood.

References

[1] S.Weinberg, "The Cosmological Constant Problem", Univ. of Texas preprint UTUTTG-12-88(1988)

[2] E.Baum, *Phys.Lett.* **133B**(1983)185

[3] S.W.Hawking, *Phys.Lett.* **134B**(1984)403

[4] S.Coleman, *Nucl.Phys.* **B310**(1988)643

[5] S.W.Hawking, *Phys.Lett.* **195B**(1987)377

[6] G.V.Lavrelashvili, V.A.Rubakov and P.G.Tinyakov, *JETP Lett.* **46**(1987)167, *Nucl.Phys.* **B299**(1988)757

[7] S.Giddings and A.Strominger, *Nucl.Phys.* **B306**(1988)890

[8] S.W.Hawking, *Phys.Rev.* **D37**(1988)904

[9] S.Coleman, *Nucl.Phys.* **B307**(1988)867

[10] S.Giddings and A.Strominger, *Nucl.Phys.* **B307**(1988)854

[11] V.A.Rubakov, *Phys.Lett.* **214B**(1988)503

[12] W.Fischler, I.Klebanov, J.Polchinski and L.Susskind, "Quantum mechanics of the googolplexus", SLAC preprint SLAC-PUB-4957 (1989)

[13] G.V.Lavrelashvili, V.A.Rubakov, M.S.Serebryakov and P.G.Tinyakov, *Nucl.Phys.* **B329**(1990)98

[14] T.Banks, Talk at Int. Seminar on Theoretical Physics, Nor-Amberd (1988), unpublished

[15] A.Strominger, *Nucl.Phys.* **B319**(1988)722

[16] A.Hosoya and M.Morikawa, *Phys.Rev.* **D39**(1989)1123

[17] M.I.Kalinin and V.N.Melnikov, Proc.VNIIFTRI **16**(1972)43 (State Committe of Standards, Moscow)

[18] J.Hartle and S.W.Hawking, *Phys.Rev.* **D28**(1983)2960

[19] A.A.Starobinsky, Talk at P.K.Sternberg State Astronomical Inst. Seminar (1983), ununpublished,
A.D.Linde, *Lett.Nuovo Cimento* **39**(1984)401

[20] V.A.Rubakov, *Phys.Lett.* **148B**(1984)280

[21] A.Vilenkin, *Phys.Rev.* **D30**(1984)549

[22] K.Kuchar, *J.Math.Phys.* **22**(1981)2640

[23] N.Caderni and M.Martellini, *Int.J.Theor.Phys.* **23**(1984)23

[24] T.Banks, *Nucl.Phys.* **B309**(1988)493

[25] P.I.Fomin, Inst.Theor.Phys., Kiev, preprint (1973)

[26] E.P.Tryon, *Nature* **246**(1973)396

[27] B.S.DeWitt, *Phys.Rev.* **160**(1967)1113

[28] C.W.Misner, in: Magic without Magic, ed.J.Klauder, San Francisco (1972).

[29] M.A.H.McCallum, in: Quantum Gravity, eds. C.J.Isham, R.Penrose and D.W.Sciama (Clarendon Press, Oxford, 1975)

[30] G.V.Lapchinsky and V.A.Rubakov, *Acta Physica Polonica* **B10**(1979)1041

[31] T.Banks, W.Fischler and L.Susskind, *Nucl.Phys.* **B262**(1985)159

[32] J.J.Halliwell and S.W.Hawking, *Phys.Rev.* **D31**(1985)1777

[33] S.Wada, *Phys.Rev.* **D34**(1986)2272

[34] A.D.Linde, *Phys.Lett.* **129B**(1983)177

COSMOLOGICAL CONSTANT, QUANTUM COSMOLOGY AND ANTHROPIC PRINCIPLE

Andrei Linde

Department of Physics
Stanford University
Stanford CA 94305-4060, USA[1]

ABSTRACT

We discuss various features of quantum cosmology in relation to the cosmological constant problem. Some ways to justify the anthropic principle in inflationary cosmology and in the baby universe theory are proposed. A possible way to create an inflationary universe in a laboratory is also discussed.

1 Introduction

The conceptual problems of quantum cosmology are very difficult. In order to solve them one should have a clear understanding of what the universe is, what the physical theory is and even what our life is. Of course, one can simply say that the universe is everything which exists, that the theory is described by the Lagrangian (or Hamiltonian) with some coupling constants, that life appears at late stages of the evolution of the universe and it cannot have any impact on the laws of Nature and on the structure of the universe. However, in the context of quantum cosmology the validity of these common sense statements becomes less clear. One may ask whether the Lagrangian existed before the universe appeared "from nothing," whether the universe existed and evolved in time before it was observed and other nasty questions which one can safely ignore in our everyday life. As will be argued below, an investigation of such questions may lead to important and rather unexpected consequences for the future development of science.

---

[1] On leave of absence from: Lebedev Physical Institute, Moscow 117924, USSR

Gravitation and Modern Cosmology
Edited by A. Zichichi *et al.*, Plenum Press, New York, 1991

One of the most difficult current problems of elementary particle theory is the cosmological constant problem. This problem was first recognized more than ten years ago, when it was understood that the effective potential $V(\phi)$ of a classical scalar field ϕ, being multiplied by $\frac{8\pi}{M_p^2}$, plays the role of the cosmological constant Λ in the Einstein equations [1]. From the comparison of the cosmological observational data with the solutions of the Einstein equations with $\Lambda \neq 0$ it follows that the observed value of the curvature scalar R at present corresponds to an extremely small value of the vacuum energy density, $V(\phi) \leq 10^{-29} g \cdot cm^{-3}$. The main problem here is to understand the reason why after a large chain of phase transitions with breaking of various symmetries in unified theories of elementary particles the vacuum energy in our asymmetric vacuum state exactly (or almost exactly) vanishes. Indeed, a most natural value of the vacuum energy density which could be expected in Kaluza-Klein theories is of the order of the Planck energy density, $\rho_P \sim M_p^4 \sim 10^{94} g \cdot cm^{-3}$, which is at least 123 orders of magnitude greater than the present vacuum energy density $V(\phi) \leq 10^{-29} g \cdot cm^{-3}$. In grand unified theories a typical value of the vacuum energy density is of the order $10^{80} - 10^{85} g \cdot cm^{-3}$, in the electroweak theory it is $O(10^{25} g \cdot cm^{-3})$. It is very difficult to understand, how all these contributions to the total value of the vacuum energy, being summed up, give us something as small as $V(\phi) \leq 10^{-29} g \cdot cm^{-3}$.

In the last year there were several interesting suggestions how to solve this problem. Here we would like to discuss the possibility to solve the cosmological constant problem by considering many universes interacting with each other only globally. Then we will discuss the baby universe theory, the possibility to create the universe in a laboratory and a possible way to justify the anthropic principle in the context of quantum cosmology.

2 A Double Universe Model

Let us first consider the model proposed in [2]. This model describes two universes, X and Y, with coordinates x_μ and y_α, respectively ($\mu, \alpha = 0, 1, \ldots, 3$) and with metrics $g_{\mu\nu}(x)$ and $\bar{g}_{\alpha\beta}(y)$, containing fields $\phi(x)$ and $\bar{\phi}(y)$ with the action of the following unusual type:

$$S = N \int d^4x d^4y \sqrt{g(x)} \sqrt{\bar{g}(y)}$$
$$\times [\frac{M_P^2}{16\pi} R(x) + L(\phi(x)) - \frac{M_P^2}{16\pi} R(y) - L(\bar{\phi}(y))] \qquad (1)$$

Here N is some normalization constant. This action is invariant under general coordinate transformations in each of the universes separately. A novel symmetry of the action is the symmetry under the transformation $\phi(x) \to \bar{\phi}(x), g_{\mu\nu}(x) \to \bar{g}_{\alpha\beta}(x)$ and under the subsequent change of the overall sign, $S \to -S$. We call this the antipodal symmetry, since it relates to each other the states with positive and negative energies.

An immediate consequence of this symmetry is the invariance under the change of the values of the effective potentials $V(\phi) \to V(\phi) + c, V(\bar{\phi}) = V(\bar{\phi}) + c$, where c is some constant. Consequently, nothing in this theory depends on the value of the effective potentials $V(\phi)$ and $V(\bar{\phi})$ in their absolute minima ϕ_0 and $\bar{\phi}_0$. (Note, that $\phi_0 = \bar{\phi}_0$ and $V(\phi_0) = V(\bar{\phi}_0)$ due to the antipodal symmetry.) This is the basic reason why it proves possible to solve the cosmological constant problem in our model.

However, our main reason to invoke this new symmetry was not just to solve the cosmological constant problem. Just as the theory of mirror particles originally was proposed in order to make the theory CP-symmetric while maintaining CP-asymmetry in its observable sector, the theory (1) is proposed in order to make the theory symmetric with respect to the choice of the sign of energy. This removes the old prejudice that, even though the overall change of sign of the Lagrangian (i.e. both of its kinetic and potential terms) does not change the solutions of the theory, one *must say* that the energy of all particles is positive. This prejudice was so strong, that several years ago people preferred to quantize *particles* with *negative energy* as *antiparticles* with *positive energy*, which caused the appearance of such meaningless concepts as negative probability. We wish to emphasize that there is no problem to perform a consistent quantization of theories which describe particles with negative energy. All difficulties appear only when there exist interacting species with both signs of energy. (Actually, this is one of the main problems of quantum cosmology, where one should quantize fields with positive energy, as well as the scalar factor a with negative energy.) In our case no such problem exists, just as there is no problem of antipodes falling down from the opposite side of the earth. The reason is that the fields $\bar\phi(y)$ do not interact with the fields $\phi(x)$, and the equations of motion for the fields $\bar\phi(y)$ are the same as for the fields $\phi(x)$ (the overall minus sign in front of $L(\bar\phi(y))$ does not change the Lagrange equations). Similarly, gravitons from different universes do not interact with each other. However, some interaction between the two universes does exist. Indeed, the Einstein equations in our case are:

$$R_{\mu\nu}(x) - \frac{1}{2}g_{\mu\nu}R(x) = -8\pi G T_{\mu\nu}(x) - g_{\mu\nu}\langle\frac{1}{2}R(y) + 8\pi G L(\bar\phi(y))\rangle, \quad (2)$$

$$R_{\alpha\beta}(y) - \frac{1}{2}\bar g_{\alpha\beta}R(y) = -8\pi G T_{\alpha\beta}(y) - \bar g_{\alpha\beta}\langle\frac{1}{2}R(x) + 8\pi G L(\phi(x))\rangle \quad (3)$$

Here $T_{\mu\nu}$ is the energy-momentum tensor of the fields $\phi(x)$, $T_{\alpha\beta}$ is the energy-momentum tensor of the fields $\bar\phi(y)$, the sign of averaging means

$$\langle R(x)\rangle = \frac{\int d^4x \sqrt{g(x)} R(x)}{\int d^4x \sqrt{g(x)}}, \quad (4)$$

$$\langle R(y)\rangle = \frac{\int d^4y \sqrt{\bar g(y)} R(y)}{\int d^4y \sqrt{\bar g(y)}}, \quad (5)$$

and similarly for $\langle L(x)\rangle$ and $\langle L(y)\rangle$. Thus, the novel feature of the theory (1) is the existence of a *global* interaction between the universes X and Y: The integral *over the whole history* of the Y-universe changes the vacuum energy density of the X-universe.

In general, the computation of the averages of the type (4), (5) may be a rather sophisticated problem. Fortunately, however, in the inflationary universe scenario [3] (at least, if the universe is not self-reproducing, see below), this task is rather trivial. Namely, the universe after inflation becomes almost flat and its lifetime becomes exponentially large. In such a case, the dominating contribution to the average values $\langle R\rangle$ and $\langle L\rangle$ comes from the late stages of the universe evolution at which the fields $\phi(x)$ and $\phi(\bar a)$ relax near the absolute minima of their effective potentials. As a result, the average value of $-L(\phi(x))$ almost exactly coincides with the value of the effective potential $V(\phi)$ in its absolute minimum at $\phi = \phi_0$, and the averaged value of the curvature scalar $R(x)$ coincides with its value at the late stages of the universe evolution, when the universe transforms to the state corresponding to the absolute minimum of $V(\phi)$. Similar results are valid for the average values of $-L(\bar\phi(y))$ and of $R(y)$ as well. In such a case one can easily show [2] that at the late stages of the universe evolution, when the fields $\phi(x)$ and $\bar\phi(y)$ relax near the absolute minima of their effective potentials, the *effective cosmological constant automatically vanishes*,

$$R(x) = -R(y) = \frac{32}{3}\pi G[V(\phi_0) - V(\bar\phi_0)] = 0 \qquad (6)$$

If the universe is self-reproducing [4], then one may encounter some difficulties when computing the averages (4), (5), since they may become infrared-divergent and the result of computation may depend on the cut-off. This question is not completely investigated yet due to the very complicated large-scale structure of the self-reproducing universe. However, one can easily avoid such questions in the theories in which $V(\phi)$ grows rapidly enough at $\phi \geq \phi^*$, since there will be no universe self-reproduction in such theories. Another problem is that the integral over d^4y in (43) renormalizes the effective Planck constant, and one should take a very small normalization factor $[N \sim exp(-\lambda^{\frac{1}{3}})$ in the theory $\lambda\phi^4]$ in order to compensate this renormalization. Another possibility is that in constructing the quantum theory in a doubled universe, one just should do it in each of the noninteracting universes separately, without an account taken of the above mentioned renormalization of N. Note also that the mechanism of the cosmological term cancellation suggested above works independently of the value of N.

The model considered above can be easily extended. For example, one can consider not only a doubled universe, but a multiuniverse as well, with the Lagrangian being a sum of different Lagrangians of different theories of different fields living in different coupled universes. Within such a theory it may be possible to justify the anthropic principle even in its most radical form.

3 Baby universes

An extremely interesting generalization of this construction was proposed by Coleman [5], Giddings and Strominger [7] and Banks [6], which is based on previous works by Hawking [8], Lavrelashvili, Rubakov and Tinyakov [9] and Giddings and Strominger [10] on the wormholes and coherence loss in quantum gravity, and also on Hawking's proposal concerning the cancellation of the cosmological term [11]. The main idea is that our universe can be split into disconnected pieces by quantum gravity effects. Baby universes created from the parent universe can carry from it an electron-positron pair, or some other combinations of particles and fields, unless this is forbidden by conservation laws. Such a process can occur in any place in our universe. Many ways were suggested to describe such a situation. The simplest one is to say that the existence of baby universes leads to a modification of the effective Hamiltonian density [5]-[7].

$$\mathcal{H}(x) = \mathcal{H}_0(\phi(x)) + \sum \mathcal{H}_i[\phi(x)]A_i \qquad (7)$$

The Hamiltonian (7) describes the fields $\phi(x)$ on the parent universe at distances much greater than the Planck scale. $\mathcal{H}_0$ is the part of the Hamiltonian which does not involve topological fluctuations. $\mathcal{H}_i(\phi)$ are some local functions of the fields ϕ, and A_i are combinations of creation and annihilation operators for the baby universes. These operators do not depend on x since the baby universes cannot carry away momentum. Coleman argues [5] that the demand of locality, on the parent universe,

$$[\mathcal{H}(x), \mathcal{H}(y)] = 0 \qquad (8)$$

for spacelike separated x and y, implies that the operators A_i must all commute. Therefore, they can be simultaneously diagonalized by the $"\alpha - states"$:

$$A_i |\alpha_i> = \alpha_i |\alpha_i> . \qquad (9)$$

If the state of the baby universe is an eigenstate of the A_i, then the net effect of the baby universes is to introduce infinite number of undetermined parameters (the α_i) into the effective Hamiltonian (7): one can just replace the operators A_i by their eigenvalues. If the universe initially is not in the A_i eigenstate, then, nevertheless, after a series of measurements the wave junction soon collapses to one of the A_i eigenstates [5]-[7].

This gives rise to an extremely interesting possibility related to the basic principles of physics. We were accustomed to believe that the main purpose of physics is to discover the Lagrangian (or Hamiltonian) of the theory which correctly describes our world. However, the question arises: if our universe did not exist sometimes in a distant past, in which sense could one speak about the existence of the laws of Nature which govern the universe? We know, for example, that the laws of our biological evolution are written in our genetic code. But where were the laws of physics written at the time when there was no universe (if there was such time)? The possible answer now is that the final structure of the (effective) Hamiltonian becomes fixed only after measurements are performed, which determine the values of coupling constants in the state in which we live with better and better precision. Different effective Hamiltonians describe different laws of physics in different (quantum) states of the universe, and by making measurements we reduce the variety of all possible laws of physics to those laws which are valid in the (classical) universe where we live.

We will not discuss this problem here any more, since it would require a thorough discussion of the difference between the orthodox (Copenhagen) and the many-world interpretation of quantum mechanics. We would like to mention only that this theory, just as the universe multiplication model considered above, opens new interesting possibilities to justify the anthropic principle.

Another important extension of the ideas of refs. [5]-[7] is connected with the possibility that the wave function of the universe depends on the values of the coupling constants in each quantum state, and it can be peaked near some particular values of these constants. A most interesting application is the possible explanation of the vanishing of the cosmological constant suggested by Coleman [12]. The main idea is closely related to the previous suggestion by Hawking [11]. According to Hawking [11], if the cosmological constant can take any given value in our universe, then the probability to find ourselves in the universe with the cosmological constant $\Lambda = \frac{8\pi}{M_P^2} V(\phi)$ would be given by

$$P(\Lambda) \sim exp(-2S_E(\Lambda)) = exp\frac{3\pi M_P^2}{\Lambda}. \quad (10)$$

In the theory discussed above the cosmological constant, like other constants, actually can take different values. However, in this theory one should not only take into account one-universe Euclidean configurations. Rather one should sum over all configurations of babies and parents (connected by Euclidean wormholes), which finally gives [12]

$$P(\Lambda) \sim exp(exp\frac{3\pi M_P^2}{\Lambda})) \quad (11)$$

Eq. (11) suggests that it is most probable to live in a quantum state of the universe with $\Lambda = 0$.

4 Top Challenge to Wormholes

Unfortunately, after two years of attempts to confirm (or disprove) eq. (11), its validity still remains unclear. Therefore at present we still do not know which of the mechanisms of the vacuum energy cancellation discussed above (if any) actually solves the cosmological constant problem. It would be very interesting to understand also how these mechanisms might work by considering some nontrivial models. Let us consider a theory which has many different absolutely stable vacuum states $\phi_1, \phi_2, ...$ with different energy densities $V(\phi_1), V(\phi_2), ...$. A simple example of such a theory is the supersymmetric $SU(5)$ model. Its effective potential $V(\Phi, H)$ has many different minima [16] with different energy density (after an account is taken of supergravity effects), which nevertheless correspond to absolutely stable vacuum states [17].

As it is shown in [4,3], in a self-reproducing inflationary universe exponentially large domains containing all possible vacua $\phi_1, \phi_2, ...$ are produced. If the difference between $V(\phi_i)$ is much smaller than M_P^4, then the number of domains with different fields ϕ_i should be of the same order of magnitude, whereas the main part of the volume of the universe would be occupied by domains with a maximal possible value of $V(\phi_i)$. In such a case one may wonder in which of these absolutely stable vacua the cosmological constant should vanish in accordance with [6,12], and why we live just in this vacuum and not in the other ones? This question is not trivial, since the choice of a particular vacuum state $|\alpha_i>$ shifts the value of Λ by some constant simultaneously in the whole universe, so it cannot make Λ vanishing in all minima of $V(\phi)$ simultaneously.

Some of the authors of the baby universe theory including Sidney Coleman do believe that the vacuum energy must vanish *in the absolute minimum* of the effective potential [13]. If this statement is true, it may provide a rather unexpected possibility to test the baby universe theory experimentally by measuring the masses of the top quark and the Higgs boson which should exist in the standard theory of electroweak interactions.

First of all, let us remember some facts about the effective potential of the Higgs field ϕ in the standard model. In the one-loop approximation [18] the effective potential is given (approximately) by

$$V = -\frac{1}{2}\mu^2\phi^2 + \frac{1}{4}\lambda\phi^4 + B\phi^4\ln(\phi/M). \tag{12}$$

Here $B = (9g^4 + 6g^2g'^2 + 3g'^2 + 9\lambda^2 - 3g_Y^4)/512\pi^2$, where g, g' and g_Y are the SU(2) gauge coupling, the U(1) gauge coupling and the top quark Yukawa coupling respectively. One can see immediately that if $B < 0$ then the vacuum is unstable. It has a maximum at some value of the field $\phi = \phi_1 > \phi_o$ and is unbounded from below at $\phi > \phi_1$. Here $\phi_o = 246$ GeV is the value of the scalar field ϕ in the local minimum of $V(\phi)$. One can easily see that this leads to the upper bound on the top quark mass $m_t = \frac{1}{\sqrt{2}}g_Y\phi_o$ [19]. To get more exact results, one should improve eq. (1) with the help of the renormalization group equation. An interesting feature of the renormalization group improved potential is that it may not be unbounded from below; a second minimum always forms at some point $\phi_2 > \phi_1$ [20]. The form of the improved potential for different values of m_t and m_H was obtained in [20] by means of computer calculations. One of the main results is that the minimum at $\phi = \phi_2$ is deeper than the minimum at $\phi = \phi_o$ for $m_t \gtrsim 95 GeV + 0.6m_H$. However, the probability of tunneling from the local minimum to the absolute one is exponentially suppressed. Therefore, unless the top quark is much heavier than $95 GeV + 0.6m_H$, the lifetime of the unstable vacuum state is greater than the age of our part of the universe by many orders of magnitude.

Of course, many people would not like even to consider the possibility that we live in an unstable vacuum state since this would be a scientific version of the doomsday prediction. However, after we learned that the universe as a whole is gravitationally unstable and that the idea of Einstein to make it static does not work, there is no reason to insist on absolute vacuum stability. Rather one should be happy if it is stable enough...

On the other hand, as we already discussed, it is very hard to understand why the vacuum energy now is vanishingly small. One may have an idea that it might be zero in the absolute minimum of the effective potential for reasons to be understood with a further development of elementary particle theory, but at the first glance it is absolutely incredible that there may exist any reason for the vacuum energy density to be smaller than $10^{-29} g.cm^{-3}$ in a *false* vacuum state in which we live now if the top quark is heavy enough.

As it was mentioned above, one may come to the same conclusion in the context of the baby universe theory. It is claimed that according to the baby universe theory the vacuum energy density should be exactly zero in a state corresponding to the *absolute* minimum of the effective potential $V(\phi)$, and that the wormhole effects do not modify the results of the standard perturbative calculations which fix the shape of $V(\phi)$ [12,13].

But this means that according to [12,13] $V(\phi_2) = 0$ for $m_t \gtrsim 95 GeV + 0.6m_H$. However, the energy difference between $V(\phi_2)$ and $V(\phi_o)$ is much bigger than $\sim 10^{25} g.cm^{-3}$. In such a case the energy density of our present unstable vacuum state $\phi = \phi_o$ would be more than 50 orders of magnitude bigger than the cosmological constraint $V(\phi_o) < 10^{-29} g.cm^{-3}$. This means that either the top quark is light enough and we live in a stable vacuum, or the theory [12] is incorrect.

Of course, we are not going to say that there is no way to generalize the baby universe theory or to suggest some other theory which may explain why the vacuum energy density should be zero even if the vacuum is false, but it is hard to imagine that this can be done without use of the anthropic principle, see below. In any case, the masses of the top quark and of the Higgs boson are still unknown, so now is the time to place one's bets !

5 Quantum Cosmology and Anthropic Principle

Until very recently, the general attitude of physicists to the anthropic principle [22], [23] was rather skeptical, to say the least. It was believed that the weak, strong and electromagnetic interactions are the same in all parts of our universe, that the fundamental constants of Nature are universal and it is meaningless to discuss the possibility of life in a universe of a different type. This attitude was somewhat changed first when it was understood that in accordance with the chaotic inflationary universe scenario the universe should contain an exponentially large (or maybe even infinitely large) number of exponentially large domains (mini-universes), in which all kinds of symmetry breaking and all kinds of compactification, which are possible in a theory of a given type, are actually realized. In other words, the universe becomes divided into many exponentially large domains with different types of low-energy physics and maybe even with different dimensionality inside each of them. This made it possible to get a justification of a version of the weak anthropic principle : if many exponentially large domains with the low-energy physics of our type do exist in the universe described by a given theory, then it is quite natural that we live in one of such domains rather than in a domain where life of our type is impossible [3]. With the invention of the theory in which many universes interact with each other only globally [2], and especially with the development of the baby universe theory [5] -[7] the situation changed even more. As will be argued below [24,25], in the context of the baby universe theory it is also possible to justify the strong anthropic principle [22] as well.

According to the baby universe theory, we may live in different quantum states of the universe $|\alpha_i>$, corresponding to different choices of the "fundamental constants," such as the vacuum energy ρ_v (cosmological constant $\Lambda = 8\pi\rho_v/M_p^2$), gravitational constant $G = M_p^{-2}$, fine structure constant α, etc. By measuring the coupling constants we actually determine which quantum state $|\alpha_i>$ we live in now, and after that we cannot "jump" to another quantum state with other constants. That is why we are used to believe that we live in a classical universe with well- determined and unchanging laws of physics.[2]

It would be very desirable to understand why we live in a universe given by a quantum state $|\alpha_i>$ with an extremely small (or even vanishing) cosmological constant, with $M_p \sim 10^{19}$ GeV, $\alpha = 1/137$, etc. As we have already mentioned, a possible idea is to compute the wave function of the universe and then to prove that it is entirely concentrated near some particular values of the coupling constants [6], [12]. However, the validity of the euclidean approach to quantum cosmology used in ref. [12] and in most of other papers on this problem is questionable. It is also quite possible also that

---

[2]An alternative possibility recently proposed in [28] is that renormalized cosmological constant in quantum cosmology becomes a function of the scale factor of the universe and rapidly disappears.

the wave function of the universe actually does not have any sharp peak at $\Lambda = 0$ [26], [27], and the possibility that it is peaked at some particular values of e, m_e or G is even much less clear. Moreover, it is not quite clear whether it is the wave function of the whole universe that is to be computed. It is known that in quantum mechanics only those statements can make sense that can be formulated in a concrete operational way. For example, at the classical level one can speak of the age of the universe t. However, the essence of the Wheeler-DeWitt equation for the wave function of the universe is that this wave function *does not depend on time*, since the total Hamiltonian of the universe, including the Hamiltonian of the gravitational field, vanishes identically [29]. Therefore if one would wish to describe the evolution of the universe with the help of its wave function, one would be in trouble. The resolution of this paradox is rather instructive [29]. The notion of evolution is not applicable to the universe as a whole since there is also no external observer with respect to the universe, and there is no external clock which does not belong to the universe. However, we do not actually ask why the universe *as a whole* evolves in the way we see it. We are just trying to understand our own experimental data. Thus, a more precisely formulated question is why *we see* the universe evolving in time in a given way. In order to answer this question one should first divide the universe into two main pieces: i) an observer with his clock and other measuring devices and ii) the rest of the universe. Then it can be shown that the wave function of the rest of the universe depends on the state of the clock of the observer, *i.e.* on his "time" [29]. This time dependence is in some sense "objective," which means that the results obtained by different (macroscopic) observers living in the same quantum state of the universe and using sufficiently good (macroscopic) measuring apparatus agree with each other [30].

This example teaches us an important lesson. By investigating the wave function of the universe *as a whole* one sometimes gets information which has no direct relevance to the observational data, *e.g.* that the universe does not evolve in time. (Moreover, it is rather difficult to give any probabilistic interpretation to the wave function of the universe as a whole, since typically it is not normalizable.) In order to describe the universe *as we see it* one should divide the universe into several macroscopic pieces and calculate a conditional probability to observe it in a given state under an obvious condition that the observer and his measuring apparatus exist. (A similar approach was used *e.g.* in ref. [31].) This simple condition, however, is very restrictive. In order to understand its meaning one should remember the famous Einstein-Podolsky-Rosen experiment, where two fermions (or photons) are produced by a decay of a spin zero particle. By measuring the spin of one of the particles an observer instantaneously determines the spin of another particle even if this particle and the observer are not in a causal contact at the moment of the measurement. This apparent paradox is resolved if one takes into account that prior to the measurement one should describe both particles by a wave function for the pair of particles rather than by separate wave functions for each of them. The knowledge of this wave function (the fact that it corresponds to a state with zero angular momentum) and the knowledge of the spin of one of these particles makes it possible to determine the spin of another particle without making any causal contact with it, simply because there is a strong correlation between spins of particles in a system with vanishing total angular momentum. For example, the conditional probability to find the second particle in a state with spin $s = -1/2$ is equal to unity under the condition that the first particle has spin $s = +1/2$.

Similarly, any scientific exploration of the properties of the universe starts with the division of the universe into two subsystems (the observer with his measuring devices and the rest of the universe). Due to the strong correlation between the properties of these two subsystems, an investigation of our own properties, the properties of our environment and our measuring devices makes it possible to say a lot about the properties of the rest of the universe such as the vacuum energy, the values of some coupling constants etc.

It is also important that in the EPR experiment one can choose different apparatus, *e.g.* the one which measures a linear polarization of the first photon, or the one which measures its circular polarization. After the measurement, the second photon will be either linearly or circularly polarized, depending on our choice of measuring device. Similarly, many kinds of life and many types of measuring devices may exist, but what we are studying is the correlation of *our* kind of life, environment and measuring devices with the properties of the rest of the universe. But this means that in the context of the baby universe theory one can actually make sense of the strong anthropic principle and try to explain why the electromagnetic coupling constant, the gravitational coupling constant and some other parameters of the theory have those particular values which we know: With other values of the coupling constants the existence of life *of our type* and of the measuring devices we use would be impossible [25], [24]. (Note, however, that in this interpretation the strong anthropic principle has nothing in common with designing the universe for the benefit of human beings or with changing the whole universe by our own will, just as the determination of the properties of the second particle in the EPR experiment and the dependence of these properties on our choice of measuring device has nothing to do with the acausal action at a distance.)

If this idea is correct, then there is no need to compute the most probable values of the cosmological constant and of the gravitational constant by investigating the wave function of the universe as a whole, and it is not surprising that some of the results of the corresponding investigation look as counterintuitive as the statement that the universe as a whole does not depend on time. Rather one should investigate the probability to measure some particular eigenvalues of the operators corresponding to the "constants" Λ, e, m_e, G in the quantum state of the universe in which we live now, under the obvious condition that we can live and perform our measurements here. In some cases (*e.g.* in the case of the constants e, m_e, G) the corresponding investigation is rather simple and straightforward. For example, according to ref. [23], an increase of m_e by more than 2.5 times would make impossible the existence of atoms. An increase of e by more than 30% would lead to an instability of protons and nuclei. A change of the strong interaction constant by more than 10 % would lead to the absence of heavy nuclei and of hydrogen in the universe. An increase or decrease of the weak interaction constant by an order of magnitude would lead to the absence of complex elements or of hydrogen respectively. An increase of the gravitational constant G by an order of magnitude would make the lifetime of the Sun so small that no biological molecules would appear on the Earth. An even bigger increase of G would lead to an extremely efficient nucleosynthesis and to the absence of hydrogen in the universe [23], [22]. A smaller value of G would slow the expansion of the universe. In such a universe the departure from thermal equilibrium during the process of baryosynthesis would be small, this process would be inefficient and the universe now would be practically empty.

This list can be easily continued, but we hope that the results discussed above clearly demonstrate that the correlation between the properties of observers and the properties of the quantum state of the rest of the universe is very strong indeed. Moreover, most of the results discussed above are not so much *anthropic*, they just show that a small deviation from the present values of coupling constants would make impossible the existence of all measuring devices we use. In the context of the baby universe theory this means that the distribution of the conditional probability to make measurements in a quantum state with given values of weak, strong, gravitational and some other "fundamental constants" proves to be sharply peaked near the actual values of these constants we have measured in the quantum state of the universe in which we live now. This peak in the probability distribution should be especially narrow in unified theories of all fundamental interactions, where the values of different coupling constants are strongly correlated due to the underlying symmetry of the theory. This indicates that if one wishes to get an order-of-magnitude estimate of the coupling constants, in some cases one can do it without computing these constants in the context of the "big fix" paradigm [12], and if one still wishes to compute them (which, of course, is a good idea), then one should compute the amplitudes corresponding to the actual experimental environment rather than the amplitudes corresponding to the metaphysical "universe in itself" investigated by an imaginary external observer. (The discussion of the time-independence of the wave function of the universe shows that in the context of quantum cosmology the subtle difference between the amplitudes of these two types can be crucially important.)

This consideration shows us a possible way to answer the question rised in the previous section. Let us consider for example the double universe model (1). One should compute the averages $\langle R(y) \rangle$ and $\langle L(y) \rangle$ over the Y-universe under the given conditions in the X-universe. If we live in a false vacuum state in the X-universe, then the main contribution to the averages $\langle R(y) \rangle$ and $\langle L(y) \rangle$ *under this condition* is also given by the false vacuum state of the theory $L(y)$, rather by the state corresponding to the absolute minimum of the effective potential $V(\bar\phi(y))$. Indeed, under the condition that the X-universe lives in a false vacuum, the effective cosmological constant in the Y-universe corresponding to the absolute minimum of the effective potential $V(\bar\phi(y))$ would be big and negative. In this case the Y-universe would be anti-de Sitter space with an extremely small lifetime, i.e. with an extremely small 4-volume. Such a state would give a negligibly small contribution to $\langle R(y) \rangle$ and $\langle L(y) \rangle$ as compared with the contribution corresponding to the false vacuum state of the theory $L(y)$, in which the effective cosmological constant vanishes (if the lifetime of the false vacuum is big enough, as is usually the case). Consequently, the effective cosmological constant should vanish in the double universe model (1) for each particular observer, whether he lives in the true vacuum state or not. This is a rather paradoxical situation: The effective cosmological constant vanishes in the absolute minimum of $V(\phi)$, for an observer who lives there, and it vanishes in the local minimum of $V(\phi)$ as well, for an observer who lives in the false vacuum state. This is the simplest example demonstrating a potential importance of the role of an observer in quantum cosmology: A naive "objective" (observer-independent) averaging would obviously lead to a big difference between the vacuum energy density in the states corresponding to a local and a global minimum of $V(\phi)$.

Perhaps, a similar situation may take place in the baby universe theory as well [21], [3]. If the mechanisms of refs. [6],[12] are operative, then the wave function of the universe probably must have not one, but several peaks at different values of Λ. One of these peaks corresponds to $\Lambda(\phi_1) = 0, \Lambda(\phi_i) \neq 0$ for $i \neq 1$, another corresponds to

$V(\phi_2) = 0, , V(\phi_i) \neq 0$ for $i \neq 2$, etc. Upon making measurements, an observer finds himself in a universe corresponding to one of these peaks. This means that he lives in a quantum state with, say, $V(\phi_2) = 0, V(\phi_1), V(\phi_3), ... \neq 0$. Any observer of our type can live only in the universe with $-10^{-29} g \cdot cm^{-3} \leq V(\phi) \leq 10^{-27} g \cdot cm^{-3}$ [37]. Since a typical difference between $V(\phi)$ in all realistic theories is many orders of magnitude bigger than $10^{-27} g \cdot cm^{-3}$, an observer of our type can find himself only in the domain of the universe containing the field $\phi = \phi_2$, independently of the value of its volume, as compared with the volume occupied by domains corresponding to other minima of $V(\phi)$.

But is it possible to solve the cosmological constant problem in the context of the baby universe theory with the help of anthropic considerations, without using ambiguous methods of computation of the wave function of the universe ? We will return to this question after discussion of another very speculative possibility: The possibility to create inflationary universe in a laboratory.

6 Hard Art of the Universe Creation

Let us examine now one more important question, relating to the possibility of creating an inflationary universe either in empty space (i.e. in vacuum) or in a laboratory. Our discussion will be based on the stochastic approach to tunneling [32,3,33]. The issue here is that quantum fluctuations in Minkowski space can bring into being an inflationary domain of size $l > H^{-1}(\phi)$, where ϕ is a scalar field produced by quantum fluctuations in this domain. The no hair theorem for de Sitter space implies that such a domain inflates in an entirely self-contained manner, independent of what occurs in the surrounding space. We could then conceive of a ceaseless process of creation of inflationary mini-universes that could take place even at the very latest stages of development of the part of the universe that surrounds us.

A description of the process whereby a region of the inflationary universe is produced as a result of quantum fluctuations could proceed in a manner similar to that for the formation of regions of the inflationary universe with a large field ϕ through the buildup of long-wave quantum fluctuations $\delta\phi$ (Brownian motion of the field ϕ). The basic difference here is that long-wave fluctuations of the massive scalar field ϕ at the time of inflation with $\phi > M_p$ are frozen in amplitude, while there is no such effect in Minkowski space. But if the build-up of quantum fluctuations $\delta\phi$ in some region of Minkowski space were to engender the creation of a fairly large and homogeneous field ϕ, then that region in and of itself could start to inflate, and such a process could stabilize (freeze in) the fluctuations $\delta\phi$ that led to its onset. In that event, one could sensibly speak of a self-consistent process of formation of inflationary domains of the universe due to quantum fluctuations in Minkowski space.

Without pretending to provide a complete description of such a process, let us attempt to estimate its probability in theories with $V(\phi) = \frac{\lambda \phi^n}{n M_p^{n-4}}$. A domain formed with a large field ϕ will only be a part of de Sitter space if $(\partial_\mu \phi)^2 \ll V(\phi)$ in its interior . This means that the size of the domain must exceed $l \sim \phi V^{-1/2}(\phi)$, and the field ϕ inside must be greater than M_p. Such a domain could arise through the build-up of quantum fluctuations $\delta\phi$ with a wavelength

$$l \sim k^{-1} \geq \phi V^{-1/2}(\phi) \sim m^{-1}(\phi). \tag{13}$$

(Note that $m^{-1}(\phi) > H^{-1}(\phi)$ during inflation.) One can estimate the dispersion $<\phi^2>_{k<m}$ of such fluctuations using the simple formula

$$<\phi^2>_{k<m} \sim \frac{1}{2\pi^2} \int_0^{m(\phi)} \frac{k^2 dk}{\sqrt{k^2 + m^2(\phi)}}$$
$$\sim \frac{m^2}{\pi^2} \sim \frac{V(\phi)}{\pi^2 \phi^2}, \tag{14}$$

and for a Gaussian distribution $P(\phi)$ for the appearance of a field ϕ which is sufficiently homogeneous on a scale $l > m^{-1}(\phi)$, one has [32]

$$P(\phi) \sim \exp(-\frac{\phi^2}{2<\phi^2>_{k<m}}) \sim \exp(-C\frac{\pi^2 \phi^4}{V(\phi)}), \tag{15}$$

where C = O(1). In particular, for a theory with $V(\phi) = \frac{\lambda}{4}\phi^4$,

$$P(\phi) \sim \exp(-C\frac{4\pi^2}{\lambda}) \tag{16}$$

Naturally, this method is rather crude; nevertheless, the estimates that it provides are quite reasonable, to order of magnitude. For example, practically the same lines of reasoning could be employed in assessing the probability of tunneling from the point $\phi = 0$ in a theory with $V(\phi) = -\frac{\lambda}{4}\phi^4$. The estimate that one obtains for the formation of a bubble of the field ϕ in this theory is also given by eq.(16). This result is in complete accord with an exact equation $P(\phi) \sim \exp(-\frac{8\pi^2}{3\lambda})$ which can be derived using Euclidean methods. In fact, one can easily verify that all results concerning tunneling in field theory and in quantum statistics at finite temperature which are discussed e.g. in [3] can be reproduced (up to a numerical factor C = O(1) in the exponent) by using the simple method suggested above. This makes the validity of the estimates (15), (16) rather plausible.

The main objection to the possibility of quantum creation of an inflationary universe in Minkowski space is that the energy conservation forbids the production of an object with positive energy out of vacuum with vanishing energy density. Within the scope of the classical field theory, in which the energy density is everywhere positive, such a process would therefore be impossible. But at the quantum level, the energy density of the vacuum is zero by virtue of the cancellation between the positive energy density of classical scalar fields, along with their quantum fluctuations, and the negative energy density associated with quantum fluctuations of fermions, or the bare negative energy of the vacuum. The creation of a positive energy-density domain through the build-up of long-wave fluctuations of the field ϕ is inevitably accompanied by formation of a region surrounding that domain in which the long-wave fluctuations of the field ϕ are suppressed, and the vacuum energy density is consequently negative. Here we are dealing with the familiar quantum fluctuations of the vacuum energy density about its zero point.

It is important that from the point of view of an external observer, the total energy of the inflationary region of the universe (and indeed the total energy of the closed inflationary universe) does not grow exponentially; the region that emerges forms a universe distinct from ours, to which it is joined only by a connecting throat (wormhole). The shortfall of long-wave fluctuations of the field ϕ surrounding the inflationary domain is quickly replenished by fluctuations arriving from neighboring regions, so the negative energy of the region near the throat can be be rapidly spread over a large volume around the inflationary domain. In such a scenario the total energy of the inflationary domain plus the energy of vacuum surrounding it will remain zero, but after the negative energy of vacuum surrounding the inflationary domain will be distributed all over the rest of the universe an observer near the inflationary domain would see it as an evaporating black hole. Indeed, it can be easily shown that the Schwarzschild radius of the inflationary domain from the point of view of an external observer is much bigger than its size $O(m^{-1})$, so it really looks like a black hole. One can show also that the time of evaporation of such a black hole is microscopically small, but it is much bigger than $m^{-1}(\phi)$. For example, it can be shown that in the theory $\frac{\lambda}{4}\phi^4$ the time of the black hole evaporation is of the order

$$t \sim \frac{\phi^3}{\lambda^{3/2} M_p^4} > (\lambda m(\phi))^{-1}.$$

This means that for a very distant observer all what happens will look as a kind of an unusual long-living quantum fluctuation, an observer at a not too big distance from the inflationary domain will see it as an evaporating black hole surrounded by space with negative vacuum energy, whereas an observer inside the inflationary domain would believe that he lives inside an inflationary universe.

Another possibility is that the negative energy density near the inflationary domain may be so large that the space around this domain will locally collapse to a singularity (like ordinary anti de Sitter space) within the time $H^{-1} \ll m^{-1}$. This singularity surrounding the inflationary domain will effectively cut it out of our space within a much smaller time than the time which would be necessary for the wormhole evaporation.

The process we are discussing here is similar to the process of the universe formation in a laboratory discussed in [34]. The difference is that we are considering formation of the universe from ordinary Minkowski vacuum, whereas in [34] formation of inflationary universe from a domain of false vacuum with large energy density was considered.

The process of the universe formation considered in [34] consists of two steps: i) formation of a small (collapsing) bubble of false vacuum and ii) its quantum tunneling after which the size of the bubble becomes bigger than H^{-1} and it begins expanding exponentially. One should note, though, that the first part of this process, which was not investigateded in [34], may be rather difficult or even impossible to realize. Indeed, the only way of producing such bubbles which we know at present is to heat some part of the universe up to the critical temperature T_c after which the phase transition to the false vacuum occurs in this domain. However, thermal energy density at the critical temperature typically is much bigger than the false vacuum energy density [3]. Such domains behave quite differently from the empty bubbles studied in [34]. Another problem is the use of the thin wall approximation in [34]. This approximation works well in the old inflationary universe scenario, but this scenario is not realistic. This approximation may work well in the extended inflation scenario, but in this scenario the theory of formation and evolution of bubbles differs from the one studied in [34]. Unfortunately, the thin wall approximation does not work at all for new and chaotic inflation.

As for the methods discussed in this section, they can be easily extended for the investigation of the process of formation of the inflationary domain in a laboratory in the new or chaotic inflationary universe scenario [35,36]. For example, one can consider the process of production of inflationary domains from the domains of space with high temperature T. To this end one should just replace expression (14) for $<\phi^2>_{k<m}$ in eq. (15) by its counterpart calculated at a nonvanishing temperature $T : <\phi^2>_{k<m} \sim \frac{Tm}{\pi^2}$:

$$P(\phi) \sim \exp(-\frac{\phi^2}{2<\phi^2>_{k<m}}) \sim \exp(-C\frac{\pi^2\phi^3}{TV^{1/2}(\phi)}).$$

We must note that a more detailed investigation is needed to prove that the process of the baby universe formation can actually occur either in the empty Minkowski space or in a laboratory. One potential difficulty is that the domain of de Sitter space produced may prove to be exponentially contracting rather than expanding. On the other hand, if it is expanding, it may tend to push away the long-wave fluctuations of the scalar field which could lead to its formation. In both these cases the part of de Sitter space formed may prove to be unstable, and it will be necessary to consider its further tunneling to a more stable field configuration [34]. Anyway, it seems that an alternative approach to the baby universe formation discussed above [32] may help us to look from another point of view on the problem of the baby universe formation. Note also, that in the eternally existing self-reproducing inflationary universe this process may occur even at the present time in those domains of the universe which are now in the inflationary phase at a density close to the Planck density. Quantum jumps of the scalar field to the Planck density space-time foam, which regularly occur in this scenario, and subsequent jumps of the field back from the space-time foam may be interpreted as a process of creation of new inflationary universes which are not attached to our universe by any regions of classical space-time. This possibility as well as other possibilities discussed above deserve further investigation.

7 Cosmological Constant Problem and Anthropic Principle

As we have seen in Section 5, anthropic considerations in the context of the baby universe theory are powerful enough to fix possible values of many coupling constants with a rather big accuracy. The possibility of explaining the small value of the vacuum energy density $|\rho_v| < 10^{-29} g \cdot cm^{-3}$ with the help of the Anthropic Principle is much less clear, but is not hopeless. From the results of ref. [37] it follows that formation of galaxies and appearance of life of our type is possible only if $-10^{-29} g \cdot cm^{-3} < \rho_v < 10^{-27} g \cdot cm^{-3}$. In the context of the baby universe theory this means that the conditional probability to measure the vacuum energy density ρ_v is concentrated in the interval between $-10^{-29} g \cdot cm^{-3}$ and $10^{-27} g \cdot cm^{-3}$ under the condition that the measurement is made by an observer of our type. Thus, the use of the Anthropic Principle in the context of the baby universe theory greatly reduces the disagreement between the theoretical expectation for the vacuum energy density and the experimental constraint $|\rho_v| < 10^{-29} g \cdot cm^{-3}$. Previously this disagreement was by 123 orders of magnitude, now the disagreement is only by a factor of $O(10^2)$, and only if the cosmological constant is positive.

The most difficult problem now is whether it is possible to get rid of the remaining factor $O(10^2)$ by anthropic considerations ? At present we do not have an entirely satisfactory answer to this question, though some possible ways towards the solution of this problem can be proposed. For example, the anthropic constraint on the vacuum energy density could be strengthened up to the desirable constraint $|\rho_v| < 10^{-29} g \cdot cm^{-3}$ if it would be possible to prove that life on the Earth has an extragalactic origin. (Any communication between galaxies would be hampered by the exponential expansion of the universe with $\rho_v > 10^{-29} g \cdot cm^{-3}$.) This could happen, for example, if the appearance of life (or of a special kind of mutations which was necessary for the appearance of humanity) was triggered by extragalactic cosmic rays. Unfortunately, at the moment we have no evidence in favor of this extravagant hypothesis.

If the top quark is heavy and we live in the false vacuum, then one may propose another anthropic argument: High-energy cosmic rays may destabilize false vacuum [38,39,33]. However, this is possible only if cosmic rays contain particles with extremely large energy. It is known, that the existence of the microwave background radiation (MBR) leads to a cut-off in the spectrum of high-energy cosmic rays, which in the final analysis might be the reason why our vacuum did not explode yet [33]. The existence of a vacuum energy density $|\rho_v| > 10^{-29} g \cdot cm^{-3}$ would decrease the density of photons in the MBR. This may remove the cut-off in the spectrum of the high-energy cosmic rays, which under certain circumstances may lead to the vacuum destabilization.

One may suggest some other, even more speculative anthropic arguments based on the investigation of the global structure of the inflationary universe [24] and on the possibility that one can create inflationary universe in a laboratory. For example, one can show that the universe with $\rho_v > 10^{-29} g \cdot cm^{-3}$ at present would be vacuum energy dominated, the distance from our galaxy to other ones would grow exponentially, and in about 10^{11} years from now there would remain no other galaxies except ours inside the horizon. This would make impossible the intergalaxy travel and the use of energy stored in other galaxies. Perhaps this would make much more difficult the process of the universe creation in the laboratory (if this process is possible at all). If one can create universes in a laboratory only under the condition that $|\rho_v| < 10^{-29} g \cdot cm^{-3}$, then one may argue that the dominant number of universes is created in the quantum state with $|\rho_v| < 10^{-29} g \cdot cm^{-3}$.

8 Conclusions

One of the main purposes of this article was to point out that the baby universe theory for the first time allows us to make sense out of the anthropic principle in application to the vacuum energy problem. This approach can explain why $-10^{-29} g \cdot cm^{-3} < \rho_v < 10^{-27} g \cdot cm^{-3}$. This still leaves without explaination the small ratio $\frac{\rho_v}{10^{-27} g \cdot cm^{-3}} \lesssim 10^{-2}$ (if ρ_v is positive), but in any theory of elementary particles we have many unexplained small numbers of this order anyway. Moreover, in the previous section we argued that there exist some ways to resolve this remaining puzzle. The three possibilities discussed in the previous section of strengthening by two orders of magnitude the anthropic constraints on the cosmological constant are obviously very premature and artificial. The main reason why we discussed them was not to propose a final solution to the cosmological constant problem but just to show that there is no "no-go" theorem on the way of completely solving the cosmological constant problem with the help of the anthropic principle.

One more issue discussed in this paper is the possibility to create the universe in a laboratory. We still do not know whether such a process is possible at all, but it seems that even if it is possible, one would have strong problems using it in any reasonable way. One cannot "pump" energy away from the new universe to ours since this would contradict energy conservation law. One cannot jump into the new universe since originally it is microscopically small and extremely dense. One even cannot send any information about himself to those people who will live in the universe which he creates. Indeed, all local properties of the universe after inflation do not depend on initial conditions at the moment of its formation: Very soon it becomes absolutely flat, homogeneous and isotropic, and any original message "imprinted" on the universe becomes unreadable.

But there exists one exception to this rule: Inflation, if it starts at a sufficiently low energy density, does not change the symmetry breaking pattern of the theory and the way of compactification of space-time. Therefore it seems that the only way to send a message to those who will live in the universe we are planning to create is to encrypt it into the properties of the vacuum state of the new universe. Such a message can be long and informative enough only if there are extremely many ways of symmetry breaking and/or patterns of compactification in the underlying theory. (This is exactly the case, e.g., in the superstring theory, which was considered for a long time as one of the main problems of this theory.) The stronger is the symmetry breaking, the more "unnatural" are relations between parameters of the theory after it, the more information the message may contain. Is it the reason we must work so hard to understand strange features of our beautiful and imperfect world ?

References

[1] A.D.Linde, JETP. Lett. **19** (1974) 183;
M. Veltman, Rockefeller University preprint (1974); Phys. Rev. Lett. **34** (1975) 77;
J. Dreilein, Phys. Rev. Lett. **33** (1974) 1243.

[2] A.D. Linde, Phys. Lett. **200B** (1988) 272.

[3] A.D. Linde, **Particle Physics and Inflationary Cosmology** (Harwood, New York, 1990);
A.D. Linde, **Inflation and Quantum Cosmology** (Academic Press, Boston, 1990).

[4] A.D. Linde, Phys. Lett. **175B** (1986) 395; Physica Scripta **T15** (1987) 169; Physics Today **40** (1987) 61.

[5] S. Coleman, Nucl. Phys. **B307** (1988) 867.

[6] T. Banks, Nucl. Phys. **B309** (1988) 493.

[7] S. Giddings and A. Strominger, Nucl. Phys. **B307** (1988) 854.

[8] S.W. Hawking, Phys. Lett. **B195** (1987) 277; Phys. Rev. **D37** (1988) 904.

[9] G.V. Lavrelashvili, V.A. Rubakov and P.G. Tinyakov, JETP Lett. **46** (1987) 167; Nucl. Phys. **B299** (1988) 757.

[10] S.B. Giddings and A. Strominger, Nucl. Phys. **B306** (1988) 890.

[11] S. Hawking, Phys. Lett. **134B** (1984) 403.

[12] S. Coleman, Nucl. Phys. **B310** (1989) 643.

[13] S. Coleman, private communication.

[14] S. Giddings and A. Strominger, Nucl. Phys. **B321** (1989) 481.

[15] G.V. Lavrelashvili, V.A. Rubakov and P.G. Tinyakov, Mod. Phys. Lett. **A3** (1988) 1231;
V.A. Rubakov and P.G. Tinyakov, Phys. Lett. **214B** (1988) 334.

[16] N.V. Dragon, Phys. Lett. **113B** (1982) 288;
P.H. Frampton and T.W. Kephart, Phys. Rev. Lett. **48** (1982) 1237;
F. Buccella, J.P. Deredinger, S. Ferrara and C.A. Savoy, Phys. Lett. **115B** (1982) 375.

[17] S. Weinberg, Phys. Rev. Lett. **48** (1982) 1776.

[18] S. Coleman and E. Weinberg, Phys. Rev. **D7** (1973) 1888.

[19] I.V. Krive and A.D. Linde, Nucl. Phys. **B117** (1976) 265; A.D. Linde, Trieste preprint IC/76/26 (1976); N.V. Krasnikov, Sov. J. Nucl. Phys. **28** (1978) 279; P.G. Hung, Phys. Rev. Lett. **42** (1979) 873; H.D. Politzer and S. Wolfram, Phys. Lett. **B82** (1979) 242; A. Anselm, JETP Lett. **29** (1979) 645; N. Cabibbo, L. Maiani, A. Parisi and R. Petronzio, Nucl. Phys. **B158** (1979) 295; A.D. Linde, Phys. Lett. **B92** (1980) 119.

[20] R. Flores and M. Sher, Phys. Rev. **D27** (1983) 1679; M. Duncan, R. Philippe and M. Sher, Phys. Lett. **B153** (1985) 165; M. Lindner, M. Sher and H.W. Zaglauer, Phys. Lett. **B228** (1989) 139; M. Sher, Phys. Reports **179** (1989) 273.

[21] V.A. Rubakov, private communication.

[22] J.D. Barrow and F.J. Tipler, **The Anthropic Cosmological Principle** (Oxford University Press, Oxford, 1986).

[23] I. L. Rosental, **Big Bang, Big Bounce: How Particles and Fields Drive Cosmic Evolution**, (Springer, Berlin, 1988).

[24] A.D. Linde, Phys. Lett. **B227** (1989) 352.

[25] V.A. Rubakov and M.E. Shaposhnikov, Mod. Phys. Lett. **4A** (1989) 107.

[26] J. Polchinski, Phys. Lett. **B219** (1989) 251.

[27] W. Fischler, I. Klebanov, J. Polchinski and L. Susskind, Nucl. Phys. **B327** (1989) 157.

[28] A. Cooper, L. Susskind and L. Thorlacius, preprint SLAC-PUB-5413 (1991).

[29] J.A. Wheeler, in: **Relativity, Groups and Topology** eds. C.M. DeWitt and J.A. Wheeler (Benjamin, New York, 1968);
B.S. DeWitt, Phys. Rev. **160** (1967) 1113.

[30] H. Everett, Rev. Mod. Phys. **29** (1957) 454; B. S. DeWitt and N. Graham, **The Many-Worlds Interpretation of Quantum Mechanics**, (Princeton University Press, Princeton, 1973).

[31] S. W. Hawking and D. N. Page, Nucl. Phys. **B264** (1986) 185 .

[32] A.S. Goncharov, A.D. Linde and V.F. Mukhanov, Int. J. Mod. Phys. **A2** (1987) 561.

[33] J. Ellis, A. Linde and M. Sher, Phys. Lett. **B252** (1990) 203.

[34] E. Farhi and A. Guth, Phys. Lett. **183B** (1987) 149;
W. Fischler, D. Morgan and J. Polchinski, Phys. Rev. **D41** (1990) 2638; Phys. Rev. **D42** (1990) 4042;
E. Farhi, A. Guth and J. Gueven, Nucl. Phys. **B339** (1990) 417.

[35] A.D. Linde, in: **Proceedings of the 1989 Summer School in High Energy Physics and Cosmology**, eds. J. Pati, S. Randjbar-Daemi, E. Sezgin and Q. Shafi (World Scientific, Singapore, 1989).

[36] A.D. Linde, Physica Scripta **T36** (1991) 30.

[37] S. Weinberg, Phys. Rev. Lett. **59** (1987) 2607; Rev. Mod. Phys. **61** (1989) 1.

[38] P. Hut and M. Rees, Nature **302** (1983) 508.

[39] M. Sher and H.W. Zaglauer, Phys. Lett. **B206** (1988) 527.

ON THE GRAVITATIONAL FIELD OF AN ARBITRARY AXISYMMETRIC MASS ENDOWED WITH MAGNETIC DIPOLE MOMENT

Igor D. Novikov[1,2] and Vladimir S. Manko[3]

[1]Nordita, Copenhagen, Denmark
[2]Astro Space Centre of P.N. Lebedev Physical Institute
  of the U.S.S.R, Academy of Sciences, Moscow, U.S.S.R.
[3]Department of Theoretical Physics, Peoples'
  Friendship University, Moscow, U.S.S.R.

1 INTRODUCTION

The well-known Birkhoff's theorem[1], the proper understanding of which became possible in many respects thanks to the paper by Peter Bergmann et al[2]. , establishes the uniqueness of the Schwarzschild spacetime as the only static spherically symmetric solution of Einstein's equations in vacuum. Interestingly, until recently it has not been known any asymptotically flat magnetostatic generalization of this very important spacetime referring to a magnetic dipole. Only in a series of papers[3-6], 70 years, after the discovery by Schwarzschild, the first exact asymptotically flat solutions of the static Einstein-Maxwell equations representing the exterior field of a massive magnetic dipole and possessing the Schwarzschild limit have been obtained by application of the nonlinear superposition technique to the Bonnor magnetic dipole solution[7] (the latter reduces to the Darmois metric[8] in the absence of magnetism). From the point of view of astrophysics, mostly dealing with deformed objects, it would be more advantageous, however, to have an asymptotically flat metric which would describe the gravitational field of an arbitrary axisymmetric mass endowed with magnetic dipole moment. In the second Section of the present article we give a possible solution of this problem by generalizing the result of the paper[3] in the case of an arbitrary set of mass-multipole moments. To obtain such generalization, we construct in the explicit form the full metric describing the nonlinear superposition of the solution[3] with an arbitrary static vacuum Weyl field. In the third Section we derive a charged version of the magnetostatic solution obtained in the second Section; the new metric which also generalizes the stationary electrovacuum solution given by the formulae (6), (7) of the paper[9] represents the field of an arbitrary axisymmetric rotating mass whose electric charge and magnetic dipole moment are defined by two independent parameters, in contradistinction to the Kerr-Newman metric[10], the magnetic field of which, being caused by the rotation of a charged source, disappears in a static limit. At the same time it should be mentioned that our electrovacuum solution, like the solutions[9], has no stationary pure vacuum limit: it becomes magnetostatic in the absence of electric charge, and electrostatic in the absence of magnetic dipole moment, describing in the latter case the field of a charged static arbitrary axisymmetric mass.

Gravitation and Modern Cosmology
Edited by A. Zichichi *et al.*, Plenum Press, New York, 1991

2 THE MAGNETOSTATIC SOLUTION

Consider the magnetostatic Einstein-Maxwell equations written in terms of the real potentials ϵ_1, ϵ_2 and γ (see [11])

$$(\epsilon_1 + \epsilon_2)\Delta\epsilon_1 = 2(\vec{\nabla}\epsilon_1)^2; \qquad (\epsilon_1 + \epsilon_2)\Delta\epsilon_2 = 2(\vec{\nabla}\epsilon_2)^2; \qquad (1)$$

$$\gamma_{,x} = \frac{4(1-y^2)}{(x^2-y^2)(\epsilon_1+\epsilon_2)^2}[x(x^2-1)\epsilon_{1,x}\epsilon_{2,x} - x(1-y^2)\epsilon_{1,y}\epsilon_{2,y} \qquad (2)$$
$$-y(x^2-1)(\epsilon_{1,x}\epsilon_{2,y} + \epsilon_{1,y}\epsilon_{2,x})];$$

$$\gamma_{,y} = \frac{4(x^2-1)}{(x^2-y^2)(\epsilon_1+\epsilon_2)^2}[y(x^2-1)\epsilon_{1,x}\epsilon_{2,x} - y(1-y^2)\epsilon_{1,y}\epsilon_{2,y} \qquad (3)$$
$$+x(1-y^2)(\epsilon_{1,x}\epsilon_{2,y} + \epsilon_{1,y}\epsilon_{2,x})],$$

where ϵ_1 and ϵ_2, depending on the prolate spheroidal coordinates x and y only, are related to the function $f(x,y)$ in the line element

$$ds^2 = k^2 f^{-1}[e^{2\gamma}(x^2-y^2)(\frac{dx^2}{x^2-1} + \frac{dy^2}{1-y^2}) + (x^2-1)(1-y^2)d\phi^2] - fdt^2, \qquad (4)$$

k being a real constant, and to the magnetic component $A_3(x,y)$ of the 4-potential of the electromagnetic field by the relations

$$\epsilon_1 = f^{1/2} + A'_3; \qquad \epsilon_2 = f^{1/2} - A'_3; \qquad (5)$$

$$A_{3,x} = k(y^2-1)f^{-1}A'_{3,y}; \qquad A_{3,y} = k(x^2-1)f^{-1}A'_{3,x}. \qquad (6)$$

In the above equations a comma denotes partial differentiation, and the operators Δ and $\vec{\nabla}$ have the form

$$\Delta \equiv k^{-2}(x^2-y^2)^{-1}\{\partial_x[(x^2-1)\partial_x] + \partial_y[(1-y^2)\partial_y]\}; \qquad (7)$$

$$\vec{\nabla} \equiv k^{-1}(x^2-y^2)^{-1/2}[\vec{x}_0(x^2-1)^{1/2}\partial_x + \vec{y}_0(1-y^2)^{1/2}\partial_y],$$

where $\vec{x}_0$ and $\vec{y}_0$ are unit vectors. The integrability condition of eqs.(2), (3) is the system (1); therefore the magnetostatic axially symmetric problem reduces to solving eqs. (1).

Using the technique developed in Ref.[11] for static Einstein-Maxwell fields we have derived the following potentials ϵ_1 and ϵ_2 satisfying (1) and representing the nonlinear superposition of the magnetostatic solution[3] with an arbitrary static vacuum Weyl field

$$\epsilon_1 = e^\psi A_+/B_+; \qquad \epsilon_2 = e^\psi A_-/B_-; \qquad (8)$$

$$A_\pm := [x^2 - y^2 + \alpha^2(x+1)^2 ab](x-1)^{1/2} \qquad (9)$$
$$\pm \alpha[(y-1)(x+y)a + (y+1)(x-y)b](x+1)^{1/2};$$

$$B_\pm := [x^2 - y^2 + \alpha^2(x-1)^2 ab](x+1)^{1/2} \qquad (10)$$
$$\pm \alpha[(y+1)(x+y)a + (y-1)(x-y)b](x-1)^{1/2},$$

where α is a real constant, ψ is any solution of Laplace's equation

$$\Delta\psi = 0, \qquad (11)$$

and the functions a and b fulfil the first-order differential equations,

$$(x-y)a_{,x} = a[(xy-1)\psi_{,x} + (1-y^2)\psi_{,y}]; \qquad (12)$$

$$(x-y)a_{,y} = b[-(x^2-1)\psi_{,x} + (xy-1)\psi_{,y}];$$
$$(x+y)b_{,x} = -b[(xy+1)\psi_{,x} + (1-y^2)\psi_{,y}];$$
$$(x+y)b_{,y} = -b[-(x^2-1)\psi_{,x} + (xy+1)\psi_{,y}].$$

It is easy to see that with $\psi = 0$, $a = b = 1$ from (8)-(10) follows the magnetic dipole solution[3], the relations (8)-(12) being its generalization in the case of an arbitrary harmonic function ψ,

From (5) and (8)-(10) one immediately finds that the metric function f of the solution (8)-(12) is given by the expressions,

$$f = (x^2-1)e^{2\psi}A^2/B^2; \tag{13}$$

$$A := [x^2 - y^2 + \alpha^2(x^2-1)ab]^2 + \alpha^2(1-y^2)[(x+y)a + (x-y)b]^2; \tag{14}$$

$$B := (x+1)[x^2 - y^2 + \alpha^2(x-1)^2ab]^2 \tag{15}$$
$$-\alpha^2(x-1)[(y+1)(x+y)a + (y-1)(x-y)b]^2.$$

On the other hand, the corresponding metric function γ and magnetic potential A_3 cannot be found that easily as the function f, since to obtain them one needs to integrate respectively eqs. (2), (3) and eqs. (6). The integration, however, can be avoided with the aid of the technique analogous to the one for the algebraic derivation of the functions γ and ω in the case of stationary vacuum solutions[12-14] accounting for the wellknown Bonnor theorem[15] which establishes the relationship between the static Einstein-Maxwell and stationary pure vacuum fields; the resulting expressions for γ and A_3 are

$$e^{2\gamma} = e^{2\gamma'}A^4(1+\alpha^2)^{-8}(x^2-y^2)^{-8} \tag{16}$$

$$A_3 = 2k\alpha e^{-\psi}A^{-1}C - 4k\alpha(1+\alpha^2)^{-1}; \tag{17}$$

$$C := (x^2-y^2)^2[(1+y)a + (1-y)b] \tag{18}$$
$$+\alpha^2(x-1)^2[(1+y)(x+y)^2a + (1-y)(x-y)^2b]ab,$$

where γ' is the function γ of the static vacuum solution determined by the potential $\psi' = (1/2)ln[(x-1)/(x+1)]+\psi$. Note, that the constants of integration in the expressions for γ and A_3 are chosen in such a way that they guarantee the asymptotic flatness of the solution (8)-(12), provided ψ, a and b have the following asymptotic behaviour when $x \to \infty$

$$\psi = O(x^{-1}); \qquad a = 1 + O(x^{-1}); \qquad b = 1 + O(x^{-1}). \tag{19}$$

To fulfil the condition (19), we choose ψ in the form of Weyl's multipoles

$$\psi = \sum_{n=1}^{\infty} \alpha_n R^{-n-1} P_n(xy/R); \qquad R \equiv (x^2+y^2-1)^{1/2}, \tag{20}$$

α_n being real constants, and $P_n()$ being the Legendre polynomials. The integration of eqs. (12) for a and b then yields

$$a = exp\left\{-\sum_{n=1}^{\infty}\sum_{l=0}^{n}\alpha_n[(x-y)R^{-l-1}P_l - 1]\right\}; \tag{21}$$

$$b = exp\left\{\sum_{n=1}^{\infty}\sum_{l=0}^{n}\alpha_n[(-1)^{n-l+1}(x+y)R^{-l-1}P_l + (-1)^n]\right\},$$

while the function γ' defined by the potential

$$\psi' = \frac{1}{2}\ln\frac{x-1}{x+1} + \sum_{n=1}^{\infty}\alpha_n R^{-n-1}P_n \qquad (22)$$

has the form[16]

$$\gamma' = \frac{1}{2}\ln\frac{x^2-1}{x^2-y^2} + \sum_{m,n=1}^{\infty}\frac{\alpha_m\alpha_n(m+1)(n+1)}{(m+n+2)R^{m+n+2}}(P_{m+1}P_{n+1} - P_m P_n) \qquad (23)$$

$$+ \sum_{n=1}^{\infty}\sum_{l=0}^{n}\alpha_n[\frac{x-y+(-1)^{n-1}(x+y)}{R^{l+1}}P_l - 1 + (-1)^{n+1}].$$

The formulae (13)-(18), (20), (21) and (23) fully determine the new magnetostatic metric which describes the gravitational field of an arbitrary static axisymmetric mass endowed with magnetic dipole moment. The metric obtained is asymptotically flat since, as one can easily verify, $f \to 1$ and $\gamma \to 0$ when $x \to \infty$. In the absence of magnetism ($\alpha = 0$) it reduces to the general static Weyl's solution $f = exp(2\psi')$, $\gamma = \gamma'$ in the form[13], ψ' and γ' being given by (22) and (23) respectively, which represents the superposition of the Schwarzschild solution with an arbitrary set of the gravitational multipoles (the parameters α_n define arbitrary static deformations of an axisymmetric mass[16]). When $\alpha_n = 0$, $\alpha \neq 0$ our solution transforms to the metric[3] for a massive magnetic dipole possessing, the Schwarzschild pure vacuum limit (mention, that one can also get a large variety of solutions with the Schwarzschild limit by simply a redefinition of the parameters $\alpha_n \to \alpha\alpha'_n$ in the general formulae).

The total mass M and magnetic dipole moment μ of our metric are given by the expressions

$$M = k(1-3\alpha^2)(1+\alpha^2)^{-1}; \qquad \mu = 8k^2\alpha^3(1+\alpha^2)^{-2}, \qquad (24)$$

and they coincide with the respective expressions for M and μ of the solution[3]; the constants α_n contribute to higher multipole moments, beginning from the mass dipole and magnetic quadrupole ones.

3 THE STATIONARY ELECTROVACUUM SOLUTION

Let us turn now to the construction of a charged generalization of the magnetostatic solution obtained in the previous Section. For this purpose we should consider already the Papapetrou line element[17],

$$ds^2 = k^2 f^{-1}[e^{2\gamma}(x^2-y^2)[\frac{dx^2}{x^2-1} + \frac{dy^2}{1-y^2}] + (x^2-1)(1-y^2)d\phi^2]$$

$$- f(dt - \omega d\phi)^2, \qquad (25)$$

in which the new unknown metric function $\omega(x,y)$ may appear due to the nonvanishing Poynting vector[18], or simply due to a spinning mass. The stationary axially symmetric electrovacuum problem then reduces to finding solutions of the system of Ernst's equations[19],

$$(Re\epsilon + \Psi\Psi^*)\Delta\epsilon = (\vec{\nabla}\epsilon + 2\Psi^*\vec{\nabla}\Psi)\vec{\nabla}\epsilon; \qquad (26)$$

$$(Re\epsilon + \Psi\Psi^*)\Delta\Psi = (\vec{\nabla}\epsilon + 2\Psi^*\vec{\nabla}\Psi)\vec{\nabla}\Psi, \qquad (27)$$

where

$$\epsilon = f - \Psi\Psi^* + i\Phi; \qquad \Psi = A_4 + iA'_3; \qquad (28)$$

$$Re\,\epsilon = f - \Psi\Psi^*; \qquad\qquad Im\,\epsilon = \Phi, \qquad(29)$$

A_4 is the electric potential, an asterisk denotes the complex conjugation, and i is the imaginary unit. The function ω is related to the potentials f, Φ and Ψ by the equations,

$$\omega_{,x} = k(y^2-1)f^{-2}[\Phi_{,y} + 2Im(\Psi^*\Psi_{,y})]; \qquad(30)$$

$$\omega_{,y} = k(x^2-1)f^{-2}[\Phi_{,x} + 2Im(\Psi^*\Psi_{,x})]. \qquad(31)$$

The metric function γ can be found once ϵ and Ψ are known[9].

To obtain the required asymptotically flat stationary electrovacuum generalization of our magnetostatic solution, we will use a particular case of the Kramer-Neugebauer transformation[20], according to which eqs. (26), (27) are satisfied by the potentials ϵ and Ψ of the form

$$\epsilon = \frac{\epsilon_0 + 2\beta\Psi_0 - \beta^2}{1 - 2\beta\Psi_0 - \beta^2\epsilon_0}; \qquad \Psi = \frac{(1+\beta^2)\Psi_0 + \beta(\epsilon_0 - 1)}{1 - 2\beta\Psi_0 - \beta^2\epsilon_0}, \qquad(32)$$

where ϵ_0 and Ψ_0 are a known solution of eqs. (32), and β is a real constant.

Let us first apply the transformation (32) to the general superposition formulae (13)-(18). From (8)-(10) and (28) we find the form of the solution (8)-(10) in terms of the potentials ϵ and Ψ satisfying eqs. (26), (27)

$$\epsilon_0 = e^{2\psi}D/B; \qquad\qquad \Psi_0 = -2i\alpha e^\psi E/B; \qquad(33)$$

$$D := (x-1)[x^2 - y^2 + \alpha^2(x+1)^2 ab]^2 \qquad(34)$$
$$-\alpha^2(x+1)[(y-1)(x+y)a + (y+1)(x-y)b]^2;$$

$$E := (x^2-y^2)^2(a-b) + \alpha^2(x^2-1)[(x+y)^2 a - (x-y)^2 b]ab, \qquad(35)$$

where B is given by (15) (one can also see that in the case of magnetostatics ϵ_0 is real and Ψ_0 is pure imaginary). Then by substituting (33)-(35) into (32) we come to a stationary electrovacuum generalization of these superposition relations, the corresponding metric function f of which, calculated from (28), (32)-(35), has the form

$$f = (1-\beta^2)^2(x^2-1)e^{2\psi}A^2/F; \qquad(36)$$

$$F := \left\{(x+1)[x^2-y^2+\alpha^2(x-1)^2 ab]^2 - \alpha^2(x-1)[(y+1)(x+y)a \right.$$
$$\left. + (y-1)(x-y)b]^2 - \beta^2 e^{2\psi}(x-1)[x^2-y^2+\alpha^2(x+1)^2 ab]^2 \qquad(37)\right.$$
$$\left. +\alpha^2\beta^2 e^{2\psi}(x+1)[(y-1)(x+y)a + (y+1)(x-y)b]^2 \right\}^2$$
$$+16\alpha^2\beta^2 e^{2\psi}\left\{(x^2-y^2)^2(a-b) + \alpha^2(x^2-1)[(x+y)^2 a - (x-y)^2 b]ab\right\}^2.$$

Since under the symmetry transformation (32) the metric function γ of the new solution is the same as γ of the of the "seed" solution ϵ_0, Ψ_0 (see[20]) the function γ corresponding to (32)-(35) is given by the formulae (14) and (16). Thus, for the construction of the whole stationary metric it only remains to find the form of the function ω by integrating eqs. (30), (31). The integration, however, turns out to be analogous to the calculation of the potential A_3 from the system of differential equations (6), and it finally leads to the following expression for ω

$$\omega = \frac{8k\alpha\beta}{(1-\beta^2)^2}\left\{\frac{e^{-\psi}C + \beta^2 e^\psi L}{A} - \frac{2(1+\beta^2)}{1+\alpha^2}\right\}; \qquad(38)$$

$$L := (x^2-y^2)^2[(1-y)a + (1+y)b] + \alpha^2(x+1)^2 \qquad(39)$$

$$\times[(1-y)(x+y)^2 a + (1+y)(x-y)^2 b] ab,$$

A and C being defined respectively by (14) and (18). Therefore, the full metric representing the charged stationary generalization of the superposition formulae (13)-(18) is constructed.

If we now again choose ψ in the form (20), then the relations (36)-(39) together with the formulae (16), (21), (23) will determine an asymptotically flat stationary metric generalizing the magnetostatic solution of the previous Section in the case of a charged spinning mass. Indeed, from (20), (23), (33)-(35), in the limit $x \to \infty$ it follows that the total mass M, charge Q, magnetic dipole moment μ and total angular momentum J are given by the expressions

$$M = \frac{k(1-3\alpha^2)(1+\beta^2)}{(1+\alpha^2)(1-\beta^2)}; \qquad Q = \frac{2k\beta(3\alpha^2-1)}{(1+\alpha^2)(1-\beta^2)}; \qquad (40)$$

$$\mu = \frac{8k^2\alpha^3(1+\beta^2)}{(1+\alpha^2)^2(1-\beta^2)}; \qquad J = \frac{16k^2\alpha^3\beta}{(1+\alpha^2)^2(1-\beta^2)}, \qquad (41)$$

whence one can see that the parameter β which defines the electric charge of the source is independent on the magnetic dipole moment determined by the parameter α. From (41) one also sees that the angular momentum J becomes zero when either electric or magnetic field disappears. Note, that the expressions (40) and (41) are the same as in the case of the solution[9], the latter being a particular case of our solution with $\alpha_n = 0$ (this is a natural result since the parameters α_n, similar to the magnetostatic case contribute to the multipole moments of higher orders).

Let us point out some particular solutions arising from the stationary metric obtained. First of all, in the absence of electric charge ($\beta = 0$) the metric reduces to the magnetostatic solution of the previous Section given by the formulae (13)-(18), (20), (21) and (23). When $\alpha = \alpha_n = 0$, $\beta \neq 0$ we come to the Reissner-Nordstrom solution[21,22] describing the field of a static spherically symmetric charged mass; therefore, the case $\alpha = 0$, $\beta \neq 0$, $\alpha_n \neq 0$ is the electrostatic generalization of the Reissner-Nordstrom solution describing the gravitational field of a charged static arbitary axisymmetric mass, the metric function f of which is defined by the expression

$$f = \frac{(1-\beta^2)^2(x^2-1)exp(2\psi)}{[(x+1)^2 - \beta^2(x-1)exp(2\psi)]^2}, \qquad (42)$$

ψ being given by (20), while the corresponding potential γ is the γ' from (23).

4 CONCLUSIONS

The metrics presented in this article are the first known asymptotically flat generalizations of the Schwarzschild spacetime in the case of an arbitrary axisymmetric mass distribution possessing a magnetic dipole moment, and their relevance to astrophysics is evident. Mention, that the general superposition formulae (8)-(18) allow one to construct a magnetic dipole generalization of any particular static vacuum metric, for which purpose it only necessary to integrate eqs. (12) for a given solution of Laplace's equation, and to choose properly the constants of integration to eliminate a magnetic charge. It should be pointed out that our solution (13)-(23) is not the only possibility to describe the gravitational field of an arbitrary axially symmetric mass endowed with magnetic dipole moment since it does not take into account the magnetic moments of higher orders. Other possibilities different from the one considered in this article will be a subject of our future publications.[1,2]

A part of this work was done when one of the authors (I. D. N.) visited NORDITA. We thank NORDITA's staff for providing excellent conditions for work.

5 REFERENCES

1. G. D. Birkhoff, Relativity and Modern Physics, Harvard Univ. Press, Cambridge, 255 (1923).

2. P. G. Bergmann, M. Cahen, and A. B. Komar, Spherically symmetric gravitational fields, *J. Math. Phys.* 6:1 (1965).

3. Ts. I. Gutsunaev, and V. S. Manko, On the gravitational field of a mass possessing a magnetic dipole moment, *Phys. Lett.* A123:215 (1987).

4. Ts. I. Gutsunaev, and V. S. Manko, New static solutions of the Einstein-Maxwell equations, *Phys. Lett.* A132:85 (1988).

5. Ts. I. Gutsunaev, V. S. Manko, and S. L. Elsgolts, New exact solutions of the static Einstein-Maxwell equations, *Class. Quant. Grav.* 6: L41 (1989).

6. V. S. Manko, New axially symmetric solutions of the Einstein-Maxwell equations, *Gen. Relat. Gravit.* 22:799 (1990).

7. W. B. Bonnor, An exact solution of the Einstein-Maxwell equations referring to a magnetic dipole, *Z. Phys.* 190:444 (1966).

8. G. Darmois, Les equations de la gravitation einsteinienne, in: "Memorial des sciences mathematique", Fasc. XXV, Gauthier-Villars, Paris (1927).

9. Ts. I. Gutsunaev, and V. S. Manko, New stationary electrovacuum generalizations of the Schwarzschild solution, *Phys. Rev.* D40:2140 (1989).

10. E. T. Newman, E. Couch, K. Chinnapared, A. Exton, A. Prakash, and R. Torrence, Metric of rotating, charged mass, *J. Math. Phys.* 6: 918 (1965).

11. Ts. I. Gutsunaev, and V. S. Manko, On a family of solutions of the Einstein-Maxwell equations, *Gen. Relat. Gravit.* 20:327 (1988).

12. M. Yamazaki, On the Hoenselaers-Kinnersley-Xanthopoulos spinning mass fields, *J. Math. Phys.* 22:133 (1981).

13. C. M. Cosgrove, Relationships between the group-theoretic and soliton- theoretic techniques for generating stationary axisymmetric gravitational solutions, *J. Math. Phys.* 21:2417 (1980).

14. W. Dietz, and C. Hoenselaers, A new class of bipolar vacuum gravitational fields, *Proc. Roy. Soc. Lond.* A382:221 (1982).

15. W. B. Bonnor, Exact solutions of the Einstein-Maxwell equations, *Z. Phys.* 161:439 (1961).

16. V. S. Manko, On the description of the external field of a static deformed mass, *Class. Quant. Grav.* 7:L209 (1990).

17. A. Papapetrou, Eine rotationssymmetrische Losung in der Allgemeinen Relativitatstheorie, *Ann. Physik* 12:309 (1953).

18. A. W. Martin, and P. L. Pritchett, Asymptotic gravitational field of the "electron", *J. Math. Phys.* 9:593 (1968).

19. F. J. Ernst, New formulation of the axially symmetric gravitational field problem II, *Phys. Rev.* 168:1415 (1968).

20. D. Kramer, and G. Neugebauer, Eine exakte stationare Losung der Einstein-Maxwell Gleichungen, *Ann. Physik* 24:59 (1969).

21. H. Reissner, Uber die Eigengravitation des electrischen Feldes nach der Einsteinschen Theorie, *Ann. Physik* 50:106 (1916).

22. G. Nordstrom, On the energy of the gravitational field in Einstein's theory, *Proc. Kon. Ned. Akad. Wet.* 20:1238 (1918).

Twistors as Spin 3/2 Charges

Roger Penrose

Mathematical Institute
Oxford, UK

Abstract It is pointed out that twistors play a role as the charges for helicity 3/2 massless fields. Since such fields can be defined consistently in general Ricci-flat 4-manifolds, a possible new approach to defining twistors in vacuum space-times is indicated.

Introduction

In this note I point out a curious and apparently significant new role for twistors, as the *charges* for massless fields of helicity 3/2. This gives another angle on a proposal that I have recently introduced (Penrose 1990) aimed at finding an appropriate twistor concept for general Ricci-flat space-times. (That approach was partly stimulated by certain ideas due to Hodges 1990a,b.)

It is perhaps appropriate, for two reasons, that I honour Peter Bergmann's 75th birthday in this way. In the first place, it was Peter's paper with Ray Sachs (Sachs and Bergmann 1958) that made clear how momentum and angular momentum play an essential role in Minkowski space-time as the corresponding charges for a massless field of spin 2, the linearized gravitational field. Secondly, this analogous role for twistors, in the case of spin 3/2, offers considerable hope that the construction that I put forward in 1975 in honour of Peter's 60th birthday (Penrose 1976), which has become known as the "non-linear graviton construction" for the (anti-)self-dual solutions of the (complex) Einstein vacuum equations, might by the use of these ideas find an appropriate generalization in which the (anti-)self-dual restriction would be removed. (See Penrose 1980 for an outline of the twistor concepts needed.)

With regard to the latter considerations, this account must be regarded as exploratory and provisional. There remain certain profound and puzzling issues yet to be resolved. Those are matters for the future. I shall therefore concentrate here largely on the situation in flat (Minkowski) space $\mathbb{M}$, though I shall also indicate what is involved in the generalization to a curved Ricci-flat space-time $\mathcal{M}$.

1 Helicity 3/2 massless fields in $\mathbb{M}$ and their charges

I use the formulation given in Penrose (1969) (see also Fierz 1940) for (classical) massless fields in $\mathbb{M}$, according to which the gauge-invariant) *field* is a valence-3 spinor $\psi_{A'B'C'}$ subject to

$$\psi_{A'B'C'} = \psi_{(A'B'C')}, \quad \nabla^{AA'}\psi_{A'B'C'} = 0. \tag{1}$$

Conventions and notation are as in Penrose and Rindler (1984, 1986); the use of primed indices here rather than unprimed ones ensures that, for a positive-frequency wavefunction, $\psi_{A'B'C'}$ describes spin 3/2 particles of positive rather than negative helicity, i.e. *helicity* 3/2.

We can, locally, find a potential $\gamma^C_{A'B'}$ for $\psi_{A'B'C'}$ subject to

$$\gamma^C_{A'B'} = \gamma^C_{(A'B')}, \quad \nabla^{AA'}\gamma^C_{A'B'} = 0 \tag{2}$$

where

$$\psi_{A'B'C'} = \nabla_{CC'}\gamma^C_{A'B'}, \tag{3}$$

the gauge freedom being

$$\gamma^C_{A'B'} \longmapsto \gamma^C_{A'B'} + \nabla^C_{B'}\nu_{A'} \tag{4}$$

where $\nu_{A'}$ satisfies the Weyl anti-neutrino equation (helicity $\frac{1}{2}$)

$$\nabla^{AA'}\nu_{A'} = 0. \tag{5}$$

We also have, locally, a "second" potential $\rho^{BC}_{A'}$ subject to

$$\rho^{BC}_{A'} = \rho^{(BC)}_{A'}, \quad \nabla^{AA'}\rho^{BC}_{A'} = 0 \tag{6}$$

where

$$\gamma^C_{A'B'} = \nabla_{BB'}\rho^{BC}_{A'} \tag{7}$$

with gauge freedom given by

$$\rho^{BC}_{A'} \longmapsto \rho^{BC}_{A'} + \nabla^C_{A'}\chi^B \tag{8}$$

when χ^B satisfies the Weyl neutrino equation (helicity $-\frac{1}{2}$)

$$\nabla_{BB'}\chi^B = 0. \tag{9}$$

Let us suppose that there is a field ψ (indices suppressed), subject to (1), in some region $\mathcal{R}$ of $\mathcal{M}$, surrounding a world-tube that we suppose contains the "sources" for ψ. Let $\mathcal{S}$ be a topological 2-sphere within $\mathcal{R}$ surrounding this source tube. We wish to obtain the charges for ψ by performing a suitable integral over $\mathcal{S}$. One procedure for doing this is to adopt the method (Penrose 1968, 1982, cf. Penrose and Rindler 1986, p75) of *spin-lowering*. For this, we select a (dual) twistor W_α, which for the present purposes means a pair of spinor fields

$$W_\alpha = (\lambda_A, \mu^{A'}) \tag{10}$$

subject to

$$\nabla_{AA'}\mu^{B'} = i\varepsilon_{A'}{}^{B'}\lambda_A, \quad \nabla_{AA'}\lambda_B = 0, \tag{11}$$

so $\mu^{B'}$, which depends linearly on position, is a solution of the (conjugate) twistor

equation
$$\nabla_A^{(A'} \mu^{B')} = 0. \tag{12}$$

Defining
$$\varphi_{A'B'} = \psi_{A'B'C'}\mu^{C'} \tag{13}$$

we find that, by virtue of (1) and (12), that φ satisfies the self-dual free Maxwell equations
$$\nabla^{AA'}\varphi_{A'B'} = 0, \quad \varphi_{A'B'} = \varphi_{(A'B')}, \tag{14}$$

and the *charge (electric $+$ i $\times$ magnetic)* that is assigns to the world-tube is given by
$$\eta = \frac{i}{4\pi}\oint \mathbf{F} \tag{15}$$

where the self-dual 2-form $\mathbf{F}$ is defined by
$$\mathbf{F} = \varphi_{A'B'} dx_A^{A'} \wedge dx^{AB'} \tag{16}$$

Note that $\mathbf{F}$ depends linearly on W_α and η depends linearly on $\mathbf{F}$. Thus, η's dependence on W_α is given by
$$\eta = Z^\alpha W_\alpha = \omega^A \lambda_A + \pi_{A'}\mu^{A'} \tag{17}$$

for some twistor
$$Z^\alpha = (\omega^A, \pi_{A'}) \tag{18}$$

describing the strength of the charge, for the field ψ, that must be assigned to the world-tube.

We see that the concept of a *twistor* Z^α indeed arises naturally here as the *charge* for a helicity 3/2 massless field. Correpondingly, a dual twistor would arise as the charge for a helicity $-3/2$ massless field. (Interchange primed and unprimed spinor indices throughout.)

Let us see how the potentials for ψ fit in with this. Note that if
$$\theta_{A'}^C = \gamma_{A'B'}^C \mu^{B'} - i\rho_{A'}^{BC}\lambda_B \tag{19}$$

then
$$\varphi_{A'B'} = \nabla_{CB'}\theta_{A'}^C, \quad \nabla_B^{A'}\theta_{A'}^C = 0, \tag{20}$$

by (3), (7), (11) and by (2), (6), (11), and defining the 1-form $\mathbf{A}$ by
$$\mathbf{A} = \theta_{BB'}dx^{BB'} \tag{21}$$

we have, from (16),
$$\mathbf{F} = 2\mathrm{d}\mathbf{A} \tag{22}$$

Note that if $\mathbf{A}$ were defined globally over $\mathscr{S}$, then the definition (15) would yield $\eta = 0$. By (19), (21), this situation always occurs if the second potential ρ, for ψ, is global over $\mathscr{S}$ (since if ρ is global then γ is certainly global), but it will also occur if merely the first potential γ is global provided that W_α belongs to the two-dimensional subspace of dual twistor space given by $\lambda_A = 0$ (see (19)). Referring to (17) we see that the latter situation — where γ is global but ρ is not necessarily so — provides us with the *factor space* of twistor space that is parameterized by the spinor $\pi_{A'}$. I shall refer to this two-dimensional complex vector space as π-*space* and the corresponding *subspace* for which $\pi_{A'} = 0$, parameterized by ω^A, as ω-*space*. We have the short exact sequence

$$0 \to \omega\text{-}space \to twistor\,space \to \pi\text{-}space \to 0 \tag{23}$$

(see Penrose and Rindler 1986, p 91) and, by the above (where "global" means global over $\mathscr{S}$):

$$\pi\text{-}space = \frac{space\ of\ global\ \psi s}{space\ of\ global\ \gamma s}, \tag{24}$$

$$\omega\text{-}space = \frac{space\ of\ global\ \gamma s}{space\ of\ global\ \rho s}. \tag{25}$$

2 Helicity 3/2 fields in Ricci-flat space-times

It was noticed many years ago by Hans Buchdahl (1958), and effectively rediscovered many times subsequently by people interested in supergravity theory etc. (eg. Deser and Zumino 1976), that the (immediate) consistency conditions for a spin 3/2 potential γ given by (2) that arise from the presence of space-time curvature, are automatically satisfied if the trace-free part of the Ricci tensor R_{ab} vanishes. Moreover, the gauge freedom is indeed given by (4), the derivative of a Weyl neutrino field ν as in (5), provided that

$$R_{ab} = 0. \tag{26}$$

It is not clear to me in what sense the description "γ-potential modulo ν-gauge" is fully *consistent* in Ricci-flat space-times, though such claims are frequently made. (In fact the equations normally used by supersymmetry theorists differ in substance and not just in appearance from those given in (2) and (5), in that the symmetry in $A'B'$ is not imposed on γ, and ν is taken to be free with $\nabla_{C(C'}\gamma^C_{A')B'}\varepsilon^{A'B'} = 0$ in addition to $\nabla^{AA'}\gamma^C_{A'B'} = 0$. However, this difference is of no consequence, the local equivalence of the different formalisms being an implication of the local solubility of the inhomogeneous Weyl neutrino equation.) The immediate consistency conditions are obtained by acting on (2) with a contracted derivative $\nabla^{B'}_A$ to obtain

$$\nabla^{B'}_A \nabla^{AA'} \gamma^C_{A'B'} = 0 \tag{27}$$

and on (4) with $\nabla^{BB'}$ to obtain

$$\nabla^{BB'} \nabla^C_{B'} \nu_{A'} = 0, \tag{28}$$

which indeed automatically vanishes when $R_{ab} = 0$, by virtue of the spinor covariant derivative commutators (Penrose and Rindler 1984) and (5). It appears to be the case that all higher-derivative consistency conditions are also automatically satisfied by virtue of (26), but we shall be seeing shortly that the "consistency" that is thereby achieved for the "γ-potential modulo ν-gauge" system is not quite so complete as might have been hoped.

Grounds for some suspicion are already contained in (24), since it would seem that one ought to be able to generalize this relation in order to define the "π-space" for any topological 2-sphere $\mathscr{S}$ in any Ricci-flat space-time $\mathscr{M}$. Although we do not have the ψ-spinors directly, because the defining equation (3) is now not invariant under the gauge transformations (4), we can still provide a concept of "global ψs". For this, we envisage a family of different γs, say $\overset{1}{\gamma},\overset{2}{\gamma},...,\overset{n}{\gamma}$, with

$$\overset{i}{\gamma} \text{ defined on } \mathcal{U}_i \tag{29}$$

where $\mathcal{U}_1, \mathcal{U}_2, ..., \mathcal{U}_n$ are open sets covering an open region $\mathcal{R}$ of $\mathcal{M}$ containing $\mathcal{S}$. On each (non-vacuous) overlap $\mathcal{U}_i \cap \mathcal{U}_j$ the corresponding γs must be gauge equivalent, i.e. there must be a family of neutrino fields $\overset{ij}{\nu}$ with

$$\overset{i}{\gamma} - \overset{j}{\gamma} = \nabla \overset{ij}{\nu} \text{ on } \mathcal{U}_i \cap \mathcal{U}_j \tag{30}$$

(spinor indices suppressed). In $\mathbb{M}$, this is indeed equivalent to having a global ψ in $\mathcal{R}$, but in $\mathcal{M}$, since ψ does not generally exist, one must resort to a piecemeal definition of this kind.

Suspicion naturally arises, however, because if this were to provide us with a definition of π-space for $\mathcal{S}$, we should also have this definition holding unchanged (free of any holonomy) as $\mathcal{S}$ is varied continuously throughout $\mathcal{R}$, and this would seem to provide a "too global" definition of π-space for $\mathcal{M}$. For numerous reasons, a π-space arising in this way is implausible. Indeed, we shall, in a moment, see more directly that fundamental difficulties confront such a global ψ.

Let us return to the situation in $\mathbb{M}$. We can consider covering $\mathcal{R}$ with just two sets $\mathcal{U}_1, \mathcal{U}_2$, where we suppose that is just a thickened out S^2 and that $\mathcal{U}_1$ and $\mathcal{U}_2$ are thickened out northern and southern hemispheres. There is an annular intersection region $\mathcal{U}_1 \cap \mathcal{U}_2$ (a thickened out equator) along which $\overset{1}{\gamma} - \overset{2}{\gamma}$ is pure gauge:

$$\overset{1}{\gamma} - \overset{2}{\gamma} = \nabla \overset{12}{\nu}. \tag{31}$$

However, if the "π-charge" is to be non-zero, we find that $\overset{12}{\nu}$ is not single-valued but it must jump by a constant spinor field as sone proceeds once around the annulus, this spinor field being a numerical multiple of (in fact $(2\pi i)^{-1} \times$) the actual value $\pi_{A'}$ of the "π-charge" of the spin 3/2 field in question (see Fig 1).

It is important to note that the reason this works is that *constant* spinors $\nu_{A'}$ do not show up in the gauge transformations (4), since then $\nabla^C_{A'}\nu_{B'} = 0$. These constants correspond to the "gauge transformations of the second kind" that are

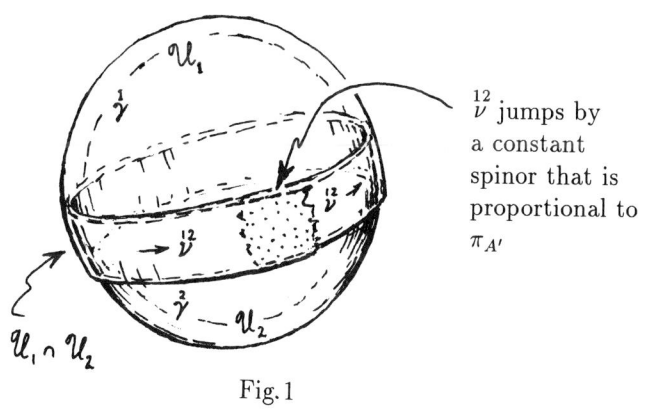

Fig. 1

familiar in electromagnetic theory. In fact the procedure just outlined is effectively an example of "evaluation" of a cohomology class, although the covering $\mathcal{U}_i$ should have been chosen more finely than this for the procedure strictly to apply: all the open sets and their various multiple intersections should be topologically trivial (unlike the annular region $\mathcal{U}_1 \cap \mathcal{U}_2$). If it is assumed that these multiple intersections are indeed topologically trivial, the evaluation procedure is as follows: we have

$$\mathbf{F} = 2d\overset{i}{\mathbf{A}}, \quad \overset{i}{\mathbf{A}} \text{ defined on } \mathcal{U}_i. \tag{32}$$

Also

$$\overset{i}{\mathbf{A}} - \overset{j}{\mathbf{A}} = d\overset{ij}{\nu}, \quad \overset{ij}{\nu} \text{ defined on } \mathcal{U}_i \cap \mathcal{U}_j; \tag{33}$$

$$\overset{ij}{\nu} + \overset{jk}{\nu} + \overset{ki}{\nu} = \overset{ijk}{\Pi} = const. \; defined \; on \; \mathcal{U}_i \cap \mathcal{U}_j \cap \mathcal{U}_k \tag{34}$$

(The fact that the $\overset{ijk}{\Pi}$ are constant follows from the above, since $d\overset{ijk}{\Pi} = 0$.) The $\overset{ijk}{\Pi}$ ($=\overset{ijk}{\Pi}_{A'}$) provide a 2-cocycle that is evaluated against the 2-cycle that is the 2-sphere $\mathscr{S}$ itself. (This amounts to adding together the different $\overset{ijk}{\Pi}$ with appropriate signs and possible redundancies.) The result is $((2\pi i)^{-1} \times)$ the required "π-charge" $\pi_{A'}$.

Suppose, now, that a slight perturbation of the space-time $\mathbb{M}$ is made so that it becomes a general curved space-time $\mathcal{M}$. The whole procedure must now break down because constant νs do not generaly exist. The π-charge, in these circumstances must necessarily turn out to be zero!

There is something very puzzling about this since it would have seemed that one "ought" to be able to piece together a family of γ-potentials for a covering of $\mathcal{R}$ in $\mathcal{M}$ that is close to the original family (with a non-zero π-charge) in $\mathbb{M}$. The above argument shows that this intuition is incorrect, however. In fact the situation is even more puzzling that just this. We can envisage a space-time $\mathcal{M}$ containing an initial flat region within which one can define a family of γ-potentials with non-zero total π-charge. As the (Ricci-flat) space-time evolves, the domain of dependence of the region $\mathcal{R}$, in which the γs are initially defined, becomes entirely within a curved region of $\mathcal{M}$, and a π-charged family of γ-potentials becomes impossible. Such a situation is illustrated in Fig 2. Here we have a thick "plate" of flat space-time that is being encroached upon by two (plane) wave-fronts, behind which are two regions of (generic) curvature. Before the waves collide, there is a region $\mathcal{R}_0 = \mathcal{U}_1 \cup \mathcal{U}_2$ (a thick shell, which is a very much thickened out 2-sphere) that has a sufficiently large part of it in the flat region, so that a non-zero π-charged pair of γs can be defined. It would *seem* that there is no obstruction to extending these γs into the whole of $\mathcal{U}_1$ and $\mathcal{U}_2$, as in Fig. 2. As the waves come together and then separate again, having passed through one another, the space-time is entirely curved; yet the domain of dependence of $\mathcal{R}_0$, though as time passes it thins down considerably from the "thick" region $\mathcal{R}_0$, is still extended enough that the π-charge can be defined — and it should be non-zero — *provided* that a sufficient family of evolved γs exists.

It should be reasonably clear that *if* the γs can be evolved adequately in this way, then we shall be presented with a contradiction. The π-charge cannot simply disappear the moment that an (arbitrarily small) amount of curvature is encountered. But

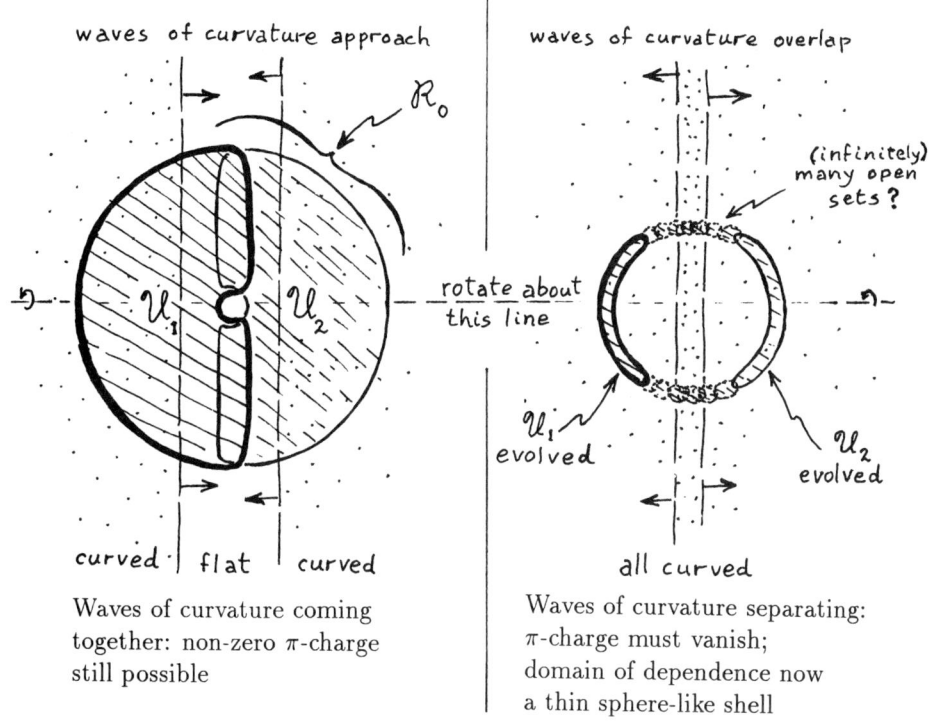

Fig. 2

what goes wrong with the γ-evolution? The answer seems to be that within the domain of dependence of each $\mathcal{U}_i(\cap \mathcal{R}_0)$ the relevant $\overset{i}{\gamma}$ does indeed evolve consistently. However, as time evolves, these domains of dependence pull away from each other and the needed intersection regions need no longer exist. One might anticipate that new open sets could be continually introduced to cover these intersections before they disappear, but these new sets, in turn, would pull away from each other. In order that this procedure can be continued sufficiently to cover the whole of the domain of dependence of $\mathcal{R}_0$, it would be necessary that the new open sets that are introduced can be chosen to be sufficiently "large". The obstruction to the size of such an open set, on which the corresponding γ can be globally defined, would be something of the nature of a "radius of convergence". However, in the situation under consideration, the space-time $\mathcal{M}$ is necessarily *non-analytic* (though it can be C^∞ and piecewise analytic), and the γs would be correpsondingly non-analytic. As the waves come together, the relevant radii of curvature may be reduced to zero and the covering ceases to exist, yielding a fundamental obstruction to the continuation of the "γ- potential modulo ν-gauge" spin 3/2 field within its own domain of dependence, despite the fact that in an *analytic* Ricci-flat space-time this description appears to be "consistent".

To some extent, the above description is merely an indirect "inference" concerning what presumably "must" happen in this puzzling and almost paradoxical situation. It would be important to gain some more direct insights into what is actually going on. (There may be some connection with the "patching problems" arising for the equations of motion of monopoles in non-Abelian gauge fields; cf. Chan, Scharbach and Tsou 1986.)

It is clear that some new insights are required if these ideas are to lead to a satisfactory "twistor" concept for general Ricci-flat space-times. it is necessary, for example, that any such concept provide a *non-linear* space of twistors, whereas the ideas, as they stand, have too much of a linear flavour about them. We need non-linearity because, in particular, we would wish to reproduce the curved twistor space of the "non-linear graviton construction" (Penrose 1976) in the case that the space-time $\mathcal{M}$ (now complex) is anti-self-dual ($\tilde{\Psi}_{A'B'C'D'} = 0$), in the notation of Penrose and Rindler 1986, p 129) and, as with that construction, we would eventually hope to be able to reconstruct the space-time geometry of a general Ricci-flat space-time from the (non-vector space) complex structure of its twistor space.

In a general way, I feel that a certain amount of optimism is in order, and to end on an encouraging note, I should point out that in a complex vacuum space time that is anti-self-dual ($\tilde{\Psi}_{A'B'C'D'} = 0$) the spin 3/2 *field* quantity $\psi_{A'B'C'}$ can be defined according to (3) and in this situation it is gauge-invariant and satisfies (1). There is now no obstruction to defining global ψ-fields, with non-zero π-charge and this is consistent with the fact that constant spinors $\Pi_{A'}$ do exist, there being a global π-space that can indeed be defined by (24). There is some hope that the twistor space defined by the α-planes in $\mathcal{M}$ (Penrose 1976) may be able to be reconstructed.

At the other end of the scale, we can consider *self*-dual Ricci-flat complex space-times ($\Psi_{ABCD} = 0$), and here it is the second potential ρ, together with its gauge quantity χ that can be consistently defined. Now it is the definition (25) that makes sense and it is the ω-space (with $\pi = 0$) that is globally defined for the space-time. To understand the *general* case (neither self-dual not anti-self-dual), however, much more work is needed.

Acknowledgments

I thank Rob Baston, Florence Tsou and particularly Andrew Hodges for stimulating discussions.

References

1. Buchdahl, H.A. (1958) On the compatibility of relativistic wave equations for particles of higher spin in the presence of a gravitational field *Nuovo Cim.* 10, 96-103

2. Chan, H.-M., Scharbach, P. and Tsou, S.T. (1986) Action principle and equations of motion for nonabelian monopoles *Annals of Physics*, 167, 454-72

3. Deser, S. and Zumino, B. (1976) Consistent supergravity *Phys. Lett.* 62B, 335-7

4. Fierz, M. (1940) Uber den Drehimpuls von Teilchen mit Ruhemasse null und beliebigen Spin *Helv. Phys. Acta.* 13, 45-60

5. Hodges, A.P. (1990a) Elemental states in *Further Advances in Twistor Theory Vol 1*: (L.J. Mason and L.P. Hughston, eds. Pitman Notes in Mathematics 231, Longman, Harlow, Essex, UK and John Wiley, New York, USA), pp 308-310 (From Twistor Newsletter 22 (1986))

6. Hodges, A.P. (1990b) Twistor diagrams and Feynman diagrams, in *Twistors in Mathematics and Physics* (T.N. Bailey and R.J. Baston, eds. Cambridge Univ. Press, Cambridge), pp 339-366

7. Penrose, R. (1968) Twistor quantization and curved space-time *Int. J. Theor. Phys.* 1, 61-99

8. Penrose, R. (1976) Nonlinear gravitons and curved twistor theory, *Gen. Rel. Grav.* 7, 31-52

9. Penrose, R. (1980) A brief outline of twistor theory, in *Cosmology and Gravitation Supergravity*, eds. P.G. Bergmann and V. de Sabbata (Plenum, New York)

10. Penrose, R. (1982) Quasi-local mass in general relativity, *Proc. Roy. Soc. London* A381, 53-63

11. Penrose, R. (1990) Twistor theory for vacuum space-times: a new approach, *Twistor Newsletter* 31, 6-8

12. Penrose, R. and Rindler, W. (1984) *Spinors and Space-Time*, Vol 1: *Two-Spinor Calculus and Relativistic Fields* (Cambridge University Press, Cambridge)

13. Penrose, R. and Rindler, W. (1986) *Spinors and Space-Time*, Vol 2: *Spinor and Twistor Methods in Space-Time Geometry* (Cambridge University Press, Cambridge)

14. Sachs, R.K. and Bergmann, P.G. (1958) Structure of particles in linearized gravitational theory, *Phys. Rev.* 112, 674-80

EXPERIMENTAL SEARCH OF GRAVITATIONAL WAVES

Guido Pizzella

Department of Physics, University of Rome
Institute for Nuclear Physics, Rome

1. Introduction

It is a great pleasure for me to be in Erice for this meeting in honour of Professor Peter Bergmann and to present the status of the experiments for the search of gravitational waves. These experiments play a crucial role for the definite experimental proof of the gravitational field theory, developed in the framework of General Relativity. This is important for integrating the gravity with the other fundamental interactions, that are based on field theories.

Our knowledge of gravity in comparison with that of the other fundamental interactions shows that, while the theory of gravity is beautifully given in the General Relativity developed by Albert Einstein, there are no experimental data that support the field character of it, namely the existence of a particle mediator between interacting bodies, the graviton. Ironically, it was the General Relativity that gave the basic idea of a field theory based on gauge invariances that express the symmetries of nature from which the fundamental forces are derived. Instead, the field character of the other forces is now well established on firm experimental bases. Although much remain to be done for the full understanding of the fermion field, the standard model gives a very good description of the boson field.

The reason for the lack of experimental data in the study of the gravitational force is due to the extreme weakness of it, about 10^{38} times weaker than the nuclear force. For this reason it is not possible to perform particle experiments of dynamical nature with the gravitational force, and one must consider the gravitational interaction of massive bodies where the other forces can be neglected, either because, being attractive and repulsive, they cancel each other, or because they have a very short range.

A very brief hystorical analysis of the gravitational wave experiments begins in 1960 with the experiment of Joe Weber at the University of Maryland. At the end of the sixties Weber reported coincidences between two room temperature antennas located at large distance one from the other one. In 1970 an experiment with cryogenic resonant antennas was started at the University of Stanford and, within one year, at the Universities of Louisiana and Rome. Soon after, also the Universities of Maryland, Perth, Rochester and Tokyo started cryogenic experiments of the same type. The search of gravitational waves with laser interferometers was born also at the University of Maryland in 1970, but it was more vigorously started at the Max-Planck Institute in Munich in 1975. Since then other groups begun a similar activity in Glasgow, MIT, Caltec, Pisa, Orsay and Tokyo. At present no clear detection of g.w. has been shown yet.

The reasons for the difficulties are, in my opinion:

a) the need of technologies that have not been fully developed yet, some of them still to be invented.

b) the need to costruct instruments that operate continuously and with the best sensitivity for very long periods of time, since the g.w. cosmic sources send signals with a very small occurrence rate.

For example, sometimes in the past resonant antennas with good sensitivity were constructed[1,2,3] but they failed to provide good data for long and continuos periods of time. The situation with the laser interferometers, about point b), seems to be also not very good.

In 1990, after 20 years of work, at least one cryogenic resonant antenna finally entered in steady operation, althought with duty cycle of roughly 50%: the antenna Explorer of the Rome group located in Geneva at CERN. It appears that within one or two years, or less, the resonant antennas of Lousiana, Perth, Stanford and Legnaro will also start to operate. This indicates that the time has finally come to perform good coincidence experiments with cryogenic resonant antennas and explore a new region of physics.

2. Gravitational wave detectors

Gravitational waves (g.w.) can be detected by measuring the change η^i of the distance between two test bodies according to the equation of the geodesic deviation

$$\frac{\delta^2 \eta^i}{\delta s^2} + u^h u^m \eta^k R^i_{mkh} = 0 \qquad [1]$$

where u^h is the four-velocity, s is the curvilinear coordinate along the geodesic line for the covariant derivative and R^i_{mkh} is the Riemann tensor.

As customary, we use now the gravitational weak field approximation, take the TT gauge, indicating with h_+ and $h_\times$ the two independent components of the gravitational wave that describe the perturbations of the metric tensor. We also choose a local Galileian system, so that the covariant derivative becomes the normal derivative and choose the coordinate system with the x axis in the travelling direction of the wave and with the origin in the location of one test mass. We obtain from [1] the following three scalar equations

$$\frac{d^2 \eta_x}{dt^2} = 0$$

$$\frac{d^2 \eta_y}{dt^2} = \frac{1}{2}(y\frac{d^2 h_+}{dt^2} + z\frac{d^2 h_\times}{dt^2}) \qquad [2]$$

$$\frac{d^2 \eta_z}{dt^2} = \frac{1}{2}(y\frac{d^2 h_\times}{dt^2} - z\frac{d^2 h_+}{dt^2})$$

We notice the transverse character of the wave gravitational field, since the relative displacement of the test bodies occurs in a plane perpendicular to the g.w. travelling direction.

The measurement of the relative displacement between the two test bodies should be done while the two bodies are in free fall, which is possible only in the outer space; one can think of two space probes or two celestial bodies like the Earth and the Moon, or two test bodies in a Spacelab environment.

In an Earth-based laboratory the two bodies must be supported against their own weight. We can classify these detectors in two categories: a) the nonresonant detectors, where the two supports for the two bodies are independent one of the other as more as possible, b) the resonant detectors, where the two bodies are linked one to the other by means of an elastic support.

2.1. The nonresonant detectors

These detectors are, essentially, Michelson or Fabry-Perot interferometers. Suppose that a g.w. travels along the direction of the arm a of the interferometer and perpendiculary to the arm b. Suppose that both arms, in the absence of g.w., have lenght l and that the g.w. has polarization h_+. From [2] we deduce that, at the g.w. arrival, the lenght of the arm a remains unchanged and the lenght of the arm b changes with time by the amount

$$\Delta l(t) = \frac{1}{2} l h_+(t) \qquad [3]$$

Therefore, we expect a small change in the interference fringes obtained by combining the light (laser) beams that travel along the two arms. From the measurement of Δl we deduce the value h_+.

There are several sources of noise which set limits to this technique. The most fundamental is the shot noise, which simulates a r.m.s. change in the distance l

$$\delta l = \sqrt{\frac{\hbar c \lambda \Delta \nu}{\pi P}} \qquad [4]$$

where P is the laser power, λ the wavelenght, $\Delta \nu$ the frequency bandwidth.

From [3] and [4] we deduce the minimum detectable value of h_+

$$h_+ \simeq 2 \frac{\Delta l}{l} = 2 \sqrt{\frac{\hbar c \lambda \Delta \nu}{\pi P l^2}} \qquad [5]$$

Putting P=100 W, λ=0.6 μm and $\Delta \nu$=1000 Hz, in order to detect a supernova in the Virgo Cluster (see later) with $h_0 \sim 3 \cdot 10^{-21}$ it is necessary to have $l \simeq 160$ km. The plans for realizing so large values of l are to construct interferometers with arms of lenght $l_0 \ll l$ and to reflect the light beams n times so that $l = n l_0$. With l_0=3 km and n=54 one obtains l=160 km.

2.2 The resonant detectors

A detector where two or more test bodies are linked together by an elastic support can be realized by means of a metallic bar[4]. Since the first pioneer experiment by Weber, the bar is usually an aluminium cylinder of lenght L and mass M. The bar must be thought in free fall during the interaction with the g.w., although it is suspended by a wire across the baricentral section in order to support it against the attraction from the Earth. We take a reference system where the centre of mass is at rest.

For this case of a continuous antenna one can follow the standard procedure to consider the displacement of a mass element in the bar under the action of the gravitational pseudoforce and of the surrounding material which acts via the Young modulus. In doing this one neglects the interaction of the gravitational force with the nuclear and atomic forces and also possible quantum effects. With this procedure one finds that the continuous antenna behaves like many elementary harmonic oscillators, each one with a resonance frequency equal to one of the various resonance frequencies of the bar. For the longitudinal fundamental mode

$$\omega_0 = \pi v / L \qquad [6]$$

where v is the sound velocity in the bar material (v = 5400 m/s in Al at the temperature of 4.2 K).

It can be shown[5] that equation [1] can be rewritten for the fundamental longitudinal mode of a cylindrical bar oriented parallel to the x axis

$$\ddot{\xi} + 2\beta_1 \dot{\xi} + \omega_0^2 \xi = \frac{2L}{\pi^2} \ddot{h} \qquad [7]$$

where ξ indicates the displacement of the bar end with respect to its equilibrium position and the Riemann tensor has been expressed in terms of the metric perturbation h that describes the g.w. (we indicate with h, for simplicity, either h_+ or $h_\times$ in the TT gauge).

The solution of eq.[7], for an impulse excitation, is

$$\xi(t) = \frac{2L}{\pi^2} \omega_0 H(\omega_0) e^{-\beta_1 t} \sin \omega_0 t \qquad [8]$$

where $H(\omega_0)$ is the Fourier component of $h(t)$ at the bar resonance ω_0. For a g.w. short burst of duration τ_g we can put, very roughly.

$$H(\omega_0) \approx h(t)\tau_g \qquad [9]$$

In the following we shall conventionally take τ_g = 0.001 s in order to express the antenna sensitivity in terms of $h(t)$ instead than $H(\omega_0)$ as it would be more appropriate.

In the framework of the detection techniques it is also convenient to express the bar vibration in terms of energy instead of displacement as indicated in [8]. The maximum total vibration energy of the bar is given by

$$E = \frac{1}{4}M\omega_0^2\xi_0^2$$

where $\xi_0 = \frac{2L}{\pi^2}\omega_0 H(\omega_0)$. It is possible to show that the energy deposited by g.w. in the fundamental longitudinal mode of a bar is

$$E = \Sigma f(\omega_0) \qquad [10]$$

where Σ is the cross section (for the best orientation)

$$\Sigma = \frac{8}{\pi}\left(\frac{v}{c}\right)^2 \frac{G}{c} M \qquad [m^2\,Hz] \qquad [11]$$

and $f(\omega_0)$ is the spectral energy density of the g.w.

$$f(\omega_0) = \frac{c^3}{8\pi G}\omega_0^2|H(\omega_0)|^2 \qquad \left[\frac{joule}{m^2 Hz}\right] \qquad [12]$$

Before discussing the g.w. antennas it is important to calculate the amplitude of the signals we expect to detect. Without any assumption on the physical mechanism of g.w. generation we consider an event taking place at a distance R from the detector, consisting in producing a g.w. burst of duration τ_g, isotropically and with total energy $M_{gw}c^2$. Then it is easy to find the corresponding value of h

$$h = \sqrt{\frac{8GM_{gw}c^2}{c^3 R^2 \omega_0^2 \tau_g}} \qquad [13]$$

at the detector location. For $\tau_g = 10^{-3}$ s, $M_{gw} = 10^{-2} M_\odot$, $\nu_0 = 1000$ Hz we find: a) for an event in the center of the Galaxy, $R = 8.5$ kpc, $h \simeq 3.5 \times 10^{-18}$. b) for an event in the Virgo cluster, $R = 10$ Mpc, $h \simeq 3 \times 10^{-21}$.

Thus our goal is to reach a sensitivity $h \leq 3 \times 10^{-21}$, so to be able to observe several events per year.

3. Optimum filtering and effective temperature

The main objective of a g.w. experiment is to detect small changes in the vibrational status of the antenna, taking into consideration both the amplitude and the phase of the oscillations. These small changes are called "energy innovations" and are produced both by external forces (say, those due to g.w.) and by the brownian and electronic noise. For measuring the antenna vibrations one uses an electromechanical transducer which we shall describe in the following section.

For a simple visualization of the problem we can consider two successive measurements of the antenna vibrational energy at the times t and $t + \Delta t$. The energy innovation due to the noise is roughly given by

$$\Delta E \cong kT \frac{\Delta t}{Q/\omega_0} + \frac{kT_n}{\beta \omega_0 \Delta t} \qquad [14]$$

where T is the bar temperature, Q its merit factor, T_n the electronic amplifier noise temperature and β the ratio of the energy that the transducer extracts from the bar to the total energy available in the bar. The first term in the right side takes into account the fact that the narrow band noise mean square displacement changes with time very slowly, that is with a time scale Q/ω_0 (say, one hour). The second term expresses the fact that the wide band noise can be reduced by averaging out over a time Δt.

The optimum filtering techniques provide a more accurate value of the minimum energy innovation that can be observed[6]. With signal to noise ratio equal to 1

$$\Delta E_{min} = 4kT_e \sqrt{\Gamma} \qquad [15]$$

with

$$T_e \cong T(1 + \frac{\beta Q T_n}{2T})$$

$$\Gamma \cong \frac{T_n}{2\beta Q T_e} \qquad [16]$$

that allows to introduce the effective temperature

$$T_{eff} = \frac{\Delta E_{min}}{k} = 4T_e \sqrt{\Gamma} \qquad [17]$$

If $\Gamma \ll 1$ then $T_{eff} \ll T$. The above formula shows the importance to have a large βQ value.

The knowledge of T_{eff} immediately gives the sensitivity in terms of $H(\omega)$. Making use of [10] and [11] we find

$$H(\omega) = \frac{L}{v^2} \sqrt{\frac{kT_{eff}}{M}} \qquad [18]$$

Finally it can be shown that the actual bandwidth $\Delta \nu$ of a resonant antenna is also related to Γ. We have [7]

$$\Delta \nu = \frac{\beta_1}{\pi \sqrt{\Gamma}} = \frac{\nu_0}{Q \sqrt{\Gamma}} \qquad [19]$$

which shows that the bandwidth is larger than the mechanical bandwidth by the factor $1/\sqrt{\Gamma}$. This is due to the fact that the resonant antenna has identical responses both to an external signal and to the brownian noise (the lorentzian curve) and the limit is set by the electronic noise. If the electronic noise were zero, the observation bandwidth were infinite.

4. The Explorer Experiment

In what follows we concentrate on the description of the Rome experiment. The first antenna of the Rome group has been named Explorer and has been installed in the CERN laboratories in Geneva. Other antennas are in preparation.

For Explorer we have used a bar of Al 5056 with lenght L=3 m and mass M=2270 kg. The bar vibrations are detected by means of a capacitive resonant transducer that consists in a resonator of Al 5056 with the shape of a mushroom [8] (fig.1); the outer diameter is 16.9825±0.0005 cm and the thickness is 0.6405±0.0005 cm. The resonator has a mechanical resonance frequency very near to that of the bar and forms, in this way, a two-coupled oscillator system.

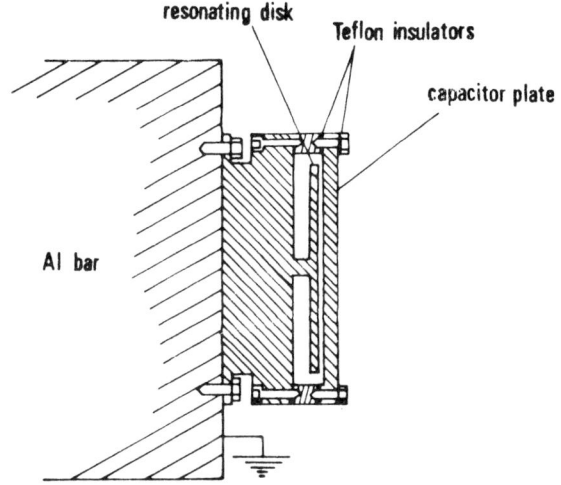

Figure 1. Scheme of the Rome resonant capacitive transducer.

The capacitor is made by placing a plate at a distance of 52±1 μm from the surface of the resonator, insulated by small washers of teflon PTFE, 100 μm thick, that is a very high quality electrical insulating material. The resonance frequency of the transducer is ν_{t0}=917.95 Hz with no applied electrical field, to be compared with the first longitudinal mode resonance frequency of the bar ν_b=915.2 Hz. By applying a bias electrical voltage V_b to the transducer its resonance frequency changes according to

$$\nu_t = \sqrt{(917.95^2 - 0.125\, V_b^2)} \qquad [20]$$

The transducer is charged by means of a battery that is then disconnected. Applying V_b =310 V it is found that the transducer loses its electrical charge at a rate of ≈ 0.1 V/day. With the above V_b we obtain $\nu_t = 911.4$ Hz that is not equal to $\nu_b = 915.2$ Hz, but it is close enough to ensure sufficient coupling. We computed that the energy transferred from the bar to the transducer in this condition is 95% of the total energy. The two coupled oscillator system has resonances

$$\nu_- = 905.1\,Hz$$

$$\nu_+ = 921.5\,Hz$$

at $V_b = 310$ V.

The transducer operates in a regime of constant electrical charge. Therefore it generates a voltage signal

$$V(t) = \alpha \xi_t(t) \qquad [21]$$

where $\xi_t(t)$ is the tranducer displacement and α (with V_b=310 V) is

$$\alpha = \gamma_t (V_b/d) C_{tr}/(C_{tr} + C_p) = 5.0 \times 10^6 \, V/m \qquad [22]$$

$\gamma_t = 0.86$ is an adimensional quantity depending on the vibration mode of the transducer disk, C_{tr}=3890 pF is the transducer capacitance and $C_p \approx 100$pF is the overall stray capacitance.

We can calculate the quantity β_c, that indicates the ratio between the energy in the transducer to the energy in the bar

$$\beta_c = \alpha^2 C_{tr}/m_t \omega^2 \qquad [23]$$

where m_t=0.35 kg is the reduced mass of the transducer. We obtain $\beta_c = 8.4 \; 10^{-3}$. However not all the energy that goes in the transducer can be used for amplifying the signal. Part of it is lost due to impedance mismatch between the transducer and the dcSQUID amplifier in spite of using a superconducting transformer as illustrated in fig. 2.

A very rough calculation for estimating this mismatching can be done as follows. From fig. 2 we estimate the current flowing in the dcSQUID coil for a given $V(t)$ generated by the transducer. The electrical resonance of the circuit of fig. 2 has frequency

$$\omega_{el} = \frac{2}{L_0 C(2 - \kappa_t^2)} = 2\pi\, 1922\, rad/s \qquad [24]$$

where L_0 is the primary inductance of the transformer, C the series of C_{tr} and C_d and $\kappa_t = 0.77$ the transformer coupling parameter. This frequency is far from the bar resonance. In this case the energy coupling parameter between the transducer and the dcSQUID can be calculated

$$\beta_L = LN^2 \kappa_t^2 C\pi\nu_R^2 = 4.4 \times 10^{-2} \qquad [25]$$

where L = 1.6 μH is the input inductance of the dcSQUID.

This gives an overall coupling parameter between bar and amplifier

$$\beta = \beta_c \cdot \beta_L = 3.6 \times 10^{-4} \qquad [26]$$

5. Measurements with the Explorer antenna

The Explorer experiment, that is located at the CERN laboratories in Geneva, started to operate at the end of June 1990 at a temperature of 3 K. Many years of work were needed to reach the planned sensitivity and, a much more difficult task, to achieve the required reliability for steady performances over long and continuous periods of time.

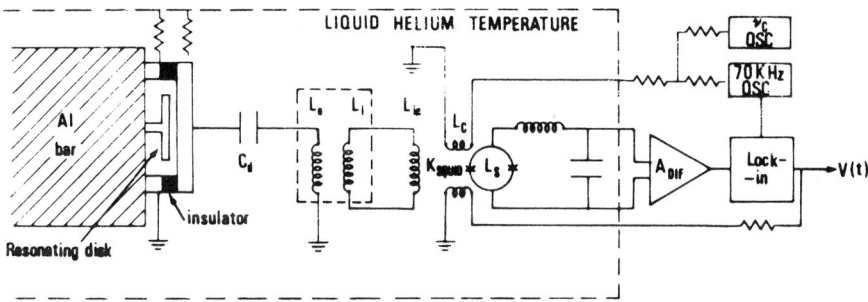

Figure 2. Scheme of the Rome experiment with the superconducting transformer and the dcSQUID amplifier.

The performance of this detector is well illustrated by fig. 3. In this figure we report the hourly averages, versus time, of the quantity h (see [18]) obtained with the following formula

$$h = H(\omega)/1000 = 2.52 \, 10^{-19} \sqrt{T_{eff}} \qquad [27]$$

with T_{eff} (hourly average) expressed in mK. This value of h (called as customary "conventional" because it is based on the assumption of a burst of g.w. with duration of 1 ms) indicates the intensity of a g.w. which would produce a signal with energy T_{eff}.

These data extend over a period of 56 days, but the detector was recording data for a total time of 28.5 days. The rest of the time, which appears in the figure as gaps between periods of data taking, was devoted to refillings with liquid nitrogen and liquid helium, and to several tests of the vacuum and electronic apparatus. In future this dead period, that amounted here to about 50% of the total time, should be reduced to about 25%, hopefully less.

Thus, in fig. 3 we have reported all the recorded data, including periods during which the detector was operating very well as well as those when some disturbances occurred. With respect to the past [9] the sensitivity appears to be definitevely better, well below the 10^{-18} threshold, except for a three day period during August when the detector was left totally unattended and the SQUID went into a less sensitive working point. Usually the SQUID working conditions are very stable; in fact the stability of the system is the real improvement in the antenna operation.

From the entire period shown in fig. 3 we have selected some periods during which the antenna was behaving particulary well. The selection criterion was to choose only those hours when the average value of T_{eff} was smaller than 15 mK. In this way a total period of 12.7 days was obtained, corresponding to about 45% of the total time of data taking.

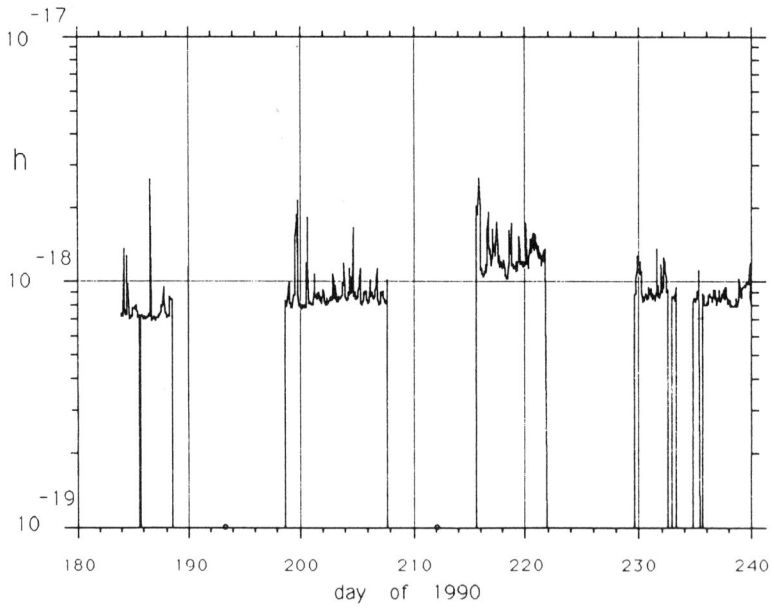

Figure 3. Hourly averages of h versus time.

For this period we plot in fig. 4 the differential distribution, with a step of 5 mK, of the data filtered with an optimum filter for the best detection of short bursts, obtaing the so-called energy innovations, which we indicate with E. Since we use a sampling time of about 0.3 s, we have in 12.7 days about $3.66 \ 10^6$ samples of energy innovations. If the only noise is that due to the brownian motion of the bar and to the electronic amplifier we expect a Boltzmann distribution with parameter equal to T_{eff}. We notice in fig. 6 a good Boltzmann distribution, with

T_{eff}=13 mK, corresponding to h=9×10^{-19},

$$exp\left(-E/13\,mK\right) \qquad [28]$$

with about three or four dozens of energy innovations E outside the expected distribution; for simplicity from now on we call these particular energy innovations "events".

It is very much likely that this "events" are due to unknown noise. Since the seismometer mounted on the Explorer cryostat does not indicate any mechanical disturbances we believe that these events are due to electromagnetic noise probably coming from the large accelerators of CERN.

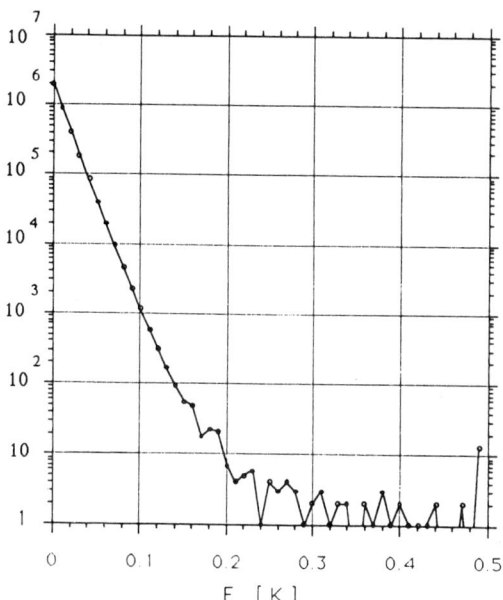

Figure 4. Differential distribution of selected data for a total period of 12.7 days out of 28.5 days.

However it is very interesting to calculate how a g.w. signal would appear on fig. 4 if it happened that a certain amount of matter M_{gw} located in the Galactic Center was converted into g.w. Using formulas [13] and [18], where in place of T_{eff} we have put the energy E of the events, we obtain

$$M_{gw}/M_o = cR^2\pi^2 k/8Gv^2\tau_g MM_o = 3.93\ 10^{-5}E \qquad [29]$$

with E is expressed in mK. Thus we see that an event with E=300 mK could be due to a g.w. produced in the Galactic Center by the conversion of about 1% of a solar mass.

This simple calculation just shows that Explorer is indeed sensitive to possible g.w.s produced in the Galaxy.

We plan to maintain Explorer in operation for a very long time, with the aim to do a coincidence experiment as soon as another antenna will become operational. When this will happen there will be a chance that a new field of physics will be born: the gravitational wave astronomy.

References

1. S.P.Boughn, W.M.Fairbank, R.P.Giffard, N.J.Hollenhorst, E.R.Mapoles, M.S.Mc Ashan, P.F.Michelson, H.J.Paik and R.C.Taber, "Observations with a low temperature, resonant mass, gravitational radiation detector", The Astrophysical Journal **261**, L19 (1982).
2. E.Amaldi, P.Bonifazi, P.Carelli, M.G.Castellano, G.Cavallari, E.Coccia, C.Cosmelli, V.Foglietti, R.Habel, I.Modena, G.V.Pallottino, G.Pizzella, P.Rapagnani and F.Ricci, "Preliminary results on the operation of a 2270 Kg cryogenic gravitational antenna with a resonant capacitive transducer and a DC SQUID amplifier", Il Nuovo Cimento **9C**, 829-845 (1986).
3. E.Amaldi, O.Aguiar, M.Bassan, P.Bonifazi, P.Carelli, M.G.Castellano, G.Cavallari, E.Coccia, C.Cosmelli,W.M.Fairbank, V.Foglietti, S.Frasca, R.Habel, W.O.Hamilton, J.Henderson, W.Johnson, M.S.McAshan, A.G.Mann, P.F.Michelson, I.Modena, B.E.Moskowitz, G.V.Pallottino, G.Pizzella, J.C.Price, P.Rapagnani, F.Ricci, N.Solomonson, T.Stevenson, R.C.Taber and B.X.Xu, "First gravity wave coincidence experiment between three resonant cryogenic detectors: Lousiana-Rome-Standford", Astronomy and Astrophysics, **216**, pag. 325-332, june 1989.
4. E.Amaldi and G.Pizzella, "The search for gravitational waves, in Relativity, Quanta and Cosmology in the Development of the Scientific Thought of Albert Einstein", Johnson Reprint Corporation, Academic Press, New York, N.Y., 1979.
5. G.Pizzella, "Gravitational-Radiation experiments", Rivista del Nuovo Cimento **5**, 369 (1975).
6. P.Bonifazi, V.Ferrari, S.Frasca, G.V.Pallottino and G.Pizzella, "Data analysis algorithms for gravitational wave experiments", Il Nuovo Cimento **1C**, 465 (1978).
7. G.V.Pallottino and G.Pizzella, "Sensitivity of a Weber type resonant antenna to monochromatic gravitational waves", Il Nuovo Cimento **7C**, 155 (1984).
8. P.Rapagnani, "Development and test at T=4.2 K of a capacitive resonant transducer for cryogenic gravitational wave antennas", Il Nuovo Cimento **5C**, 385 (1982).
9. E.Amaldi, P.Bonifazi, P.Carelli, M.G.Castellano, G.Cavallari, E.Coccia, C.Cosmelli, V.Foglietti, S.Frasca, R.Habel, I.Modena, R.Onofrio, G.V.Pallottino, G.Pizzella, P.Rapagnani and F.Ricci, "Operation of the 2270 kg g.w. resonant antenna of the Rome group", Proc. of 13th Texas Symposium on Relativistic Astrophysics, Ed. M.P.Ulmer, World Scientific, pag. 18-19 (1987).

A SIMPLE MODEL OF THE UNIVERSE WITHOUT SINGULARITIES

Nathan Rosen

Department of Physics, Technion–Israel Institute of Technology
Haifa 32000, Israel

Mark Israelit

School of Education of the Kibbutz Movement, University of Haifa
Oranim, Tivon, Israel

1. INTRODUCTION

For a long time the "standard" model of the universe [1] was popular among physicists. However, this model presents certain difficulties, such as the flatness problem, the homogeneity and isotropy problem, and the horizon (or causality) problem [2]. To overcome them the standard model has been modified by the addition of inflationary scenarios, generally based on theories the validity of which has not yet been established [2,3]. However, even with these modifications one difficulty remains, the initial singularity, or big bang, of the standard model. It is hoped that the introduction of the quantum theory to describe the early behavior of the universe will remove this singularity, but so far no satisfactory way of doing this has been found.

Recently, a simple cosmological model was proposed [4] which is free from the difficulties of the standard model. The purpose of the present work is to discuss some aspects of this model.

2. ASSUMPTIONS AND CONSEQUENCES

It is assumed that the universe is homogeneous and isotropic and that its geometry is described by the Robertson-Walker metric,

$$ds^2 = dt^2 - R^2(t)[(1 - kr^2)^{-1}dr^2 + r^2 d\theta^2 + r^2 sin^2\theta d\varphi^2], \tag{1}$$

where R is the scale factor of the universe and $k = \pm 1, 0$. With matter density $\rho(t)$ and pressure $P(t)$, the Einstein field equations,

$$G^\nu_\mu = -8\pi T^\nu_\mu, \tag{2}$$

give the relations

$$\dot{R}^2 = (8\pi/3)\rho R^2 - k, \tag{3}$$

$$\ddot{R} = -(4\pi/3)R(\rho + 3P), \qquad (4)$$

with a dot denoting a time derivative. From the Bianchi identities one gets the energy equation

$$\dot{\rho} + 3(\dot{R}/R)(\rho + P) = 0, \qquad (5)$$

with ρ and P related by an equation of state.

The model considered is assumed to be free from singularities and to oscillate in time. If we take $t = 0$ when R has its minimum value, so that $\dot{R} = 0$, $\ddot{R} > 0$, then we see from Eq. (3) that we must have $k = +1$. From Eq. (4) we see that, at $t = 0$, P must be negative and $-P > \rho/3$.

With $k = +1$, the universe is closed, and R can be considered to be its radius. The universe has the geometry of a three-dimensional spherical surface, and its volume is $2\pi^2 R^3$.

It will be assumed that there exists a limiting value for the density of matter just as there exists a limiting value c for the velocity of matter. From the fundamental constants c, G, and $\hbar$ one can form the Planck density

$$\rho_{pc} = c^5/\hbar G^2 = 5.16 \times 10^{93} \text{ g cm}^{-3}, \qquad (6)$$

or, in general relativity units,

$$\rho_p = c^3/\hbar G = 3.83 \times 10^{65} \text{cm}^{-2}. \qquad (7)$$

We now assume that the limiting density is of the order of magnitude of the Planck density [5,6,7]. For the sake of concreteness let us take it to be *equal* to the Planck density. At such an extremely high density we have no knowledge about the properties of matter. We can imagine that in this limit state the matter properties are intimately associated with those of space-time in the sense that the matter tensor is covariantly related to the metric tensor [8],

$$T_{\mu\nu} = \rho_p g_{\mu\nu}, \qquad T^\nu_\mu = \rho_p \delta^\nu_\mu, \qquad (8)$$

so that for $\rho = \rho_p$ we have

$$P = -\rho. \qquad (9)$$

The limit state can be given another interpretation. It is seen that having Eq. (8) holding is equivalent to having a cosmological constant in the Einstein equations, so that instead of (2), we have

$$G_{\mu\nu} + \Lambda g_{\mu\nu} = -8\pi T_{\mu\nu}, \qquad (10)$$

with $\Lambda = 8\pi \rho_p$ and, in the present case, $T_{\mu\nu} = 0$.

Let us assume that the universe began its expansion (at $t = 0$) in this state, which we call the "prematter" state. The field equations give

$$R = R_I \cosh(t/R_I), \qquad (11)$$

with $R_I = (3/8\pi \rho_p)^{1/2} = 5.58 \times 10^{-34}$ cm, the initial value of R. As the expansion proceeds, ρ decreases, and the prematter goes over into ordinary matter at a high temperature. One can think of the prematter as a continuum which goes over into matter consisting of particles as ρ decreases. Which particles? Presumably, at the beginning the most fundamental particles, those which subsequently combine to form quarks and leptons. This hot matter behaves like isotropic radiation having the equation of state

$$P = \frac{1}{3}\rho. \tag{12}$$

To describe the transition from the prematter period to the radiation-dominated period we need an equation of state that goes over into (9) and (12) under suitable conditions. Many such equations are possible. As an example one might assume the simplest one given by

$$P = \frac{1}{3}\rho(1 - 4\rho/\rho_p). \tag{13}$$

Substituting into (5), one obtains

$$\rho = a^4 \rho_p/(a^4 + R^4) \quad (a = \text{const}). \tag{14}$$

One can characterize the prematter period by $R \ll a$ and the radiation period by $R \gg a$. From somewhat lengthy considerations involving the radiation period, the present (dust-dominated) state of the universe, and the transition from one to the other [4], one estimates that (roughly) $a = 1 \times 10^{-3}$cm. This value will be assumed hereafter.

Substituting (14) into (13), we get

$$P = a^4 \rho_p (\frac{1}{3}R^4 - a^4)/(a^4 + R^4)^2. \tag{15}$$

With (14) and (15), Eqs. (3) and (4) can be written

$$\dot{R}^2 = (\beta R^2 - a^4 - R^4)/(a^4 + R^4), \tag{16}$$

$$\ddot{R} = \beta R(a^4 - R^4)/(a^4 + R^4)^2, \tag{17}$$

where

$$\beta = (8\pi/3)a^4 \rho_p = 3.21 \times 10^{54} \text{cm}^2. \tag{18}$$

3. EARLY HISTORY OF THE UNIVERSE

It is convenient to divide the early history of the universe into the following periods:

a) The prematter period ($R \ll a$)

From Eqs. (14) and (15) one has $\rho = \rho_p$ and $P = -\rho_p$. The behavior of R is given by (11). These relations hold to a good approximation even if $R \sim a/10$. For example, for $t = 1.27 \times 10^{-42}$ sec ($= 3.81 \times 10^{-32}$cm), Eq. (11) gives $R/R_I = 2.25 \times 10^{29}$, and $R = 1.26 \times 10^{-4}$ cm. Except at the very beginning, Eq. (11) describes an inflationary expansion.

b) The prematter-radiation transition period ($R \sim a$)

In this case, since β as given by (18) is very large compared to R^2 and a^2, Eq. (16) can be integrated to give

$$t = t_0 + R_I \int_{R_0}^{R} (1 + u/a)^{\frac{1}{2}} u^{-1} du, \tag{19}$$

where t_0 and R_0 are any pair of corresponding values of t and R in the prematter period.

It should be noted from (17) that, for $R = a$, $\ddot{R} = 0$, so that the expansion rate $\dot{R}$ has its maximum value. Since, for this value of R, β as given by (18) is much larger than R^2, we see that (16) gives for this maximum value

$$\dot{R}_{max} = (\beta/2a^2)^{\frac{1}{2}} = 1.27 \times 10^{30}. \tag{20}$$

From (15) one sees that for $R^4 = 3a^4$ one has $P = 0$, so that, for larger values of R, P is positive. One finds that P has its maximum value for $R^4 = 7a^4$. It might be mentioned in passing that for $R^4 = 3a^4$ the mass of the universe, $2\pi^2 \rho R^3$, has its maximum value, which is 4.31×10^{57} cm, or 5.81×10^{85} g.

c) The radiation period ($R \gg a$)

From Eqs. (14) and (15) one sees that now

$$\rho = a^4 \rho_p / R^4, \tag{21}$$

and the equation of state (12) is valid. From Eq. (16), fixing the integration constant with the help of (19), one gets as the relation between R and t

$$R^2 = 2\beta^{\frac{1}{2}} t - t^2. \tag{22}$$

Once one gets into the radiation period the present model agrees with the standard model.

4. THERMODYNAMICS OF THE EARLY UNIVERSE

Let us consider the temperature T and the entropy S of the universe during the prematter and radiation periods. Assuming $\rho = \rho(T)$, $P = P(T)$, and equilibrium, we write

$$dS(V,T) = \frac{1}{T}[d(\rho V) + PdV], \tag{23}$$

where V is the volume of the closed universe,

$$V = 2\pi^2 R^3. \tag{24}$$

From the integrability condition for Eq. (23) one finds

$$\frac{dP}{dT} = \frac{1}{T}(\rho + P), \tag{25}$$

and the integral of (23) is given by

$$S = \frac{V}{T}(\rho + P), \tag{26}$$

where we have taken the constant of integration to vanish.

From Eqs. (5) and (24) we get the energy relation

$$\frac{d}{dt}(\rho V) = -P \frac{dV}{dt}, \tag{27}$$

which, with the help of (25), can be written

$$\frac{dS}{dt} = 0, \tag{28}$$

with S given by (26), so that entropy is conserved.

Making use of Eq. (13) in (25) and integrating, one gets

$$\rho(1 - \frac{\rho}{\rho_p})^7 = \sigma T^4, \tag{29}$$

where we have taken the integration constant σ equal to the Stefan-Boltzmann constant, $\sigma = 6.24 \times 10^{-64}$ cm^{-2}(°K)$^{-4}$, so as to get the Stefan-Boltzmann law for $\rho \ll \rho_p$. With ρ given by Eq. (14), we get

$$T = (\frac{\rho_p}{\sigma})^{\frac{1}{4}} \frac{aR^7}{(a^4 + R^4)^2}, \tag{30}$$

where $(\rho_p/\sigma)^{\frac{1}{4}} = 1.574 \times 10^{32}$ °K. For $R = R_I$ one finds that the initial temperature is

$$T_I = 2.65 \times 10^{-180} \text{ °K}, \tag{31}$$

so that at $t = 0$ the universe is extremely cold.

From (30) one finds that T has its maximum value for $R^4 = 7a^4$, so that T and P have their maxima for the same value of R. For this value of R Eq.(30) gives $T_{max} = 7.41 \times 10^{31}$ °K, a considerable change from the initial temperature.

5. OVERVIEW

According to the present model, the behavior of the universe can be described as follows:

The universe begins its expansion from a very cold, homogeneous and isotropic "cosmic egg" in the form of a three-dimensional spherical surface with radius R_I. The initial density ρ_I, according to (14), is very close to the Planck density ρ_p, so that $(\rho_p - \rho_I)/\rho_p < 10^{-121}$. The universe is in the prematter state, characterized by an enormous tension, $-P = \rho$. The initial rate of expansion $\dot{R}_I = 0$, but the tension produces a large initial acceleration, $\ddot{R}_I = 1.79 \times 10^{33}$ cm^{-1}. After a time $t = 1.2 \times 10^{-42}$ sec the radius of the universe has increased by a factor 3.6×10^{28} to a value $R = 2 \times 10^{-5}$ cm, and the expansion rate has grown to $\dot{R} \simeq 3 \times 10^{28}$, while the density and tension have remained practically constant, although the temperature has increased to 1.28×10^{20} °K.

At about $t = 1.2 \times 10^{-42}$ sec one can consider the prematter-radiation transition period to begin. During this period radical changes take place. At the time $t = 1.31 \times 10^{-42}$ sec, when $R = a = 1 \times 10^{-3}$ cm, $\rho = \frac{1}{2}\rho_p$, and $P = -\frac{1}{6}\rho_p$, one has $\ddot{R} = 0$, so that the expansion rate has its maximum, $\dot{R} = 1.27 \times 10^{30}$. Thereafter deceleration sets in - gravitational attraction begins to brake the expansion. Very soon after this inflection point, at $t = 1.32 \times 10^{-42}$ sec, when $R^4 = 3a^4$ and $\rho = \frac{1}{4}\rho_p$, one has a pressure $P = 0$; it becomes positive after that and grows rapidly. At $t = 1.33 \times 10^{-42}$ sec, when $R^4 = 7a^4$. P and T have their maximum values: $P = \frac{1}{6}\rho = \frac{1}{48}\rho_p$, and $T = 7.41 \times 10^{31}$ °K. At $t = 5 \times 10^{-42}$ sec, with $R = 2 \times 10^{-2}$ cm, one has $P = \frac{1}{3}\rho$, so that the radiation era has begun. At this moment $\rho = 2.35 \times 10^{60}$ cm^{-2}, the Stefan-Boltzman radiation temperature is 7.8×10^{30} °K and the mass is $m = 3.78 \times 10^{56}$ cm $= 5.10 \times 10^{84}$ g. From this time on the universe behaves as if it had started from the big bang.

The radiation-dominated period and the subsequent matter-dominated period proceed as in the standard model [1]. To describe the present state is difficult because there is considerable uncertainty in the present density ρ_N arising from the existence of dark matter in the universe. Let us take the present Hubble constant $H_N = 5 \times 10^{-29}$ cm^{-1}, so that the critical density $\rho_c = (3/8\pi)H_N^2 = 2.98 \times 10^{-58}$ cm^{-2}. In the model under discussion [4] it is assumed arbitrarily that the present radius $R_N = 2.5/H_N$. One then finds that $\rho_N = 1.16\rho_c = 3.46 \times 10^{-58}$ cm^{-2}, the age of the universe is $t_N = 1.37 \times 10^{10} y$, and the present mass of the universe $m_N = 8.54 \times 10^{29}$ cm $= 1.15 \times 10^{58}$ g. It follows that the maximum radius $R_{max} = 7.25 R_N = 3.63 \times 10^{29}$ cm and will be reached at a time $t_{max} = 6.03 \times 10^{11} y$. At this time $\dot{R} = 0, \ddot{R} = -1.38 \times 10^{-30}$ cm^{-1}, and the universe begins to contract to its initial state as part of its oscillatory behavior.

It should be stressed that there is a great deal of arbitrariness in the numerical values given, due to the arbitrary choice of the equation of state (13) and the various parameters used. However, the purpose of this work is not to obtain a specific model of the universe, but to show that a simple non-singular model is possible.

In the model described here there are none of the difficulties usually discussed. At $t = 0$ the density was nearly equal to the limiting density ρ_p, so that the universe was nearly homogeneous and isotropic. Hence one can understand why it remained nearly homogeneous and isotropic in the course of time.

With nearly identical conditions at all points for $t = 0$ and therefore for all subsequent times, the question of causal relations between distant points becomes unimportant. One has nearly the same conditions in the neighborhoods of two distant point, not because there is communication between them, but because their present behavior was determined by nearly the same conditions in the past.

The so-called flatness problem is no problem in the present model. One can discuss the question of flatness with the help of Eq. (3). Whenever $\dot{R}^2 >> 1$ (or $\rho R^2 >> 1$), one can neglect k on the right side of the equation, so that the situation is as if $k = 0$, i.e., as if the universe is flat, even though we have $k = 1$. Only near $t = 0$, i.e., for $t < 10^{-43}$ sec, was the curvature parameter important, and in our matter-dominated period it is becoming important as R is approaching R_{max}. In between, k was negligible and the universe behaved as if it were flat.

In conclusion, we wish Peter many more years of his fruitful work, which has contributed so much to the advancement of physics.

REFERENCES

1. S. Weinberg, Gravitation and Cosmology (Wiley, New York, 1972).

2. A.H. Guth, Phys. Rev. D **23** (1981) 347.

3. A.D. Linde, Rep. Prog. Phys. **47** (1984) 925.

4. M. Israelit and N. Rosen, Astrophys. J. **342** (1989) 627.

5. E.B. Gliner, Sov. Phys. - Doklady **15** (1970) 559.

6. M.A. Markov, Sov. Phys. - JETP Lett. **36** (1982) 265.

7. N. Rosen, Astrophys. J. **297** (1985) 347.

8. E.B. Gliner, Sov. Phys. - JETP **22** (1966) 378.

STRING THEORY AND THE QUANTIZATION OF GRAVITY

N. Sánchez

Observatoire de Paris
Section de Meudon, Demirm
92195 Meudon Principal Cedex, France

INTRODUCTION

Perhaps the main challenge in theoretical physics today is the unification of all interactions including gravity. At present, string theories appear as the best candidates to achieve such an unification. However, several technical and conceptual problems remain and a quantum theory of gravity is still non-existent. Continuous effort over the last quarter of a century has demonstrated the many difficulties encountered in repeated attempts to construct such a theory and has also indicated some of the particular properties which an eventual complete theory will have to posses. The amount of work in that direction can be by now presented in two different sets which have most evolved (and remain) separated: (i) conceptual unification (introduction of the uncertainty principle in general relativity, the interpretation problem and the concept of 'observables', Q.F.T. in curved space time and by accelerated observers, Hawking radiation and its consequences, the Wheeler-De Witt equation and the "wave function of the universe"...) (ii) grand unification (the unification of all interactions including gravity from the particle physics point of view, in which, gravity is considered as a massless spin two particle (the graviton), such as in supergravities, Kaluza-Klein theories and the more succefull: superstrings).
Most of the work in the part (i) ("conceptual unification") treats gravity in the context of point particle field theory, that is what we call conventional quantum gravity. Few are the works in such part, which have incorporated the novelty of strings. On the other hand, most of the work done on strings do not treat the connection with the main problems of quantum gravity. (The main motivation and the impact of modern string theory is to give a consistent quantum theory of gravity, but, unfortunately, most of the work done on strings do not address to this problem).

Gravitation and Modern Cosmology
Edited by A. Zichichi *et al.*, Plenum Press, New York, 1991

Whatever the final theory of the world will be, if it is to be a theory of everything, we would like to know what new understanding it will give us about the singularities of classical general relativity. If string theory would provide a theory of quantum gravity, it should give us a proper theory (not yet existent) for describing the ultimate state of quantum black holes and the initial (very early) state of the universe. That is, a theory describing the physics (and the geometry) at Planck energies and lengths.

A QUANTUM THEORY OF GRAVITY MUST BE FINITE

Many attempts have been done to quantize gravity. The problem most often discussed in this connection is the one of renormalizability of Einstein theory (or its various generalizations) when quantized as a local quantum field theory. Actually, even deeper conceptual problems arise when ones tries to combine quantum concepts with General Relativity. Let us begin by an argument showing conceptually that a consistent quantum theory of gravity must be *finite*. In other words, it is not possible to conceive a renormalizable Q.FT. when the gravitational interaction is included.

What is a renormalizable Q.F.T.? This is a theory with some domain of validity characterized by energies E such that $E < \Lambda$. Here, the scale Λ is characteristic of the model under consideration: (e.g., $\Lambda=1$ Gev for QED, 100 Gev for the standard model or 10^{16} Gev for GUT,etc. One always applies the QFT in question till finite energy (or zero distance) for virtual processes and finds usually ultra-violet infinities. These divergences reflect the fact that the model is unphysical for energies $\Lambda << E << \infty$. In a renormalizable QFT these infinities can be absorbed in a finite number (usually few) parameters like coupling constants and mass ratios, which are not predicted by the model in question. One would need a more general theory valid at energies beyond Λ in order to compute these renormalized parameters (presumably from others more fundamental). For example, M_W/M_Z is calculable in a Grand Unified Theory, whereas it must be fitted to its experimental value in the standard electro-weak model.

Now, what about quantization of gravity ? The relevant energy scale is the Planck mass ($M_{Planck} \sim 10^{19}$ Gev). At this mass, the Schwarzschild radius (r_s) equals the Compton wavelength (λ_c) of a particle. Then, if we imagine particles heavier than M_{Planck}, their size λ_c will be smaller than r_s. In other words, to localize them in a region of size λ_c will be in conflict with what we know about the Schwarzschild radius from general relativity. Such heavy objects ($M_{Planck} \sim 10^{-5}$ g) can not behave (if they really exist) as usual point particles do in relativistic QFT. This means that M_{Planck}

gives the order of magnitude for the heaviest point particles. There cannot be point particles beyond M_{Planck} in a relativistic QFT as soon as gravity is included. This shows that we can not conceive a renormalizable QFT including gravity since there can not exist a theory at energies higher than M_{Planck} whose ignorance is responsible for the infinities of quantum gravity. (If ultraviolet divergences appear in a quantum theory of gravity, there is no way to interpret them as coming from a higher energy scale as it is usually done in QFT). Hence, a consistent theory including gravity must be *finite*. All dimensionless physical quantities must be computable in it. These conceptual arguments are consistent with all failed attempts to construct renormalizable field theories of quantum gravity.

Another consequence of these arguments is the following: Since a quantum theory of gravitation would describe the highest possible mass scale, such model must also include all other interactions in order to be consistent and true. That is, one may ignore higher energy phenomena in a low energy theory, but the opposite is not true. To give an example, a theoretical prediction for graviton-graviton scattering at energies of the order of M_{Planck} must include *all* particles produced in a real experiment. That is, in practice, all existing particles in nature, since gravity couples to *all matter*.

These simple arguments, based on the renormalization group [1] lead us to an important conclusion: a consistent quantum theory of gravitation must be a *theory of everything* (TOE). So rich a theory should be very complicated to find and to solve. In particular, it needs the understanding of the present desert beween 1 and 10^{16} Gev. There is an additional dimensional argument about the inference that a Quantum theory of gravity implies a TOE. There exist only three fundamental physical magnitudes: length, time and energy and hence three fundamental dimensional constants: c, h and G. All other parameters being dimensionless, they must be calculable in a unified quantum theory including gravity, and therefore, a theory like this must be a TOE. From the purely theoretical side the only serious candidate at present for a TOE is string theory. Unfortunately, most of the research work done on strings consider the strings in Minkowski space-time. All string models exhibit particle spectra formed by tower of massive particles going up to infinite mass and hence passing by M_{Planck}. If these states are to be considered as point particles, one arrives, for masses larger than M_{Planck}, to the clash between general relativity and quantum mechanics described above. A solution of this paradox could be that the particle spectrum of string models is at energies $E \gg M_{Planck}$ very different from what we know today on the basis of perturbation theory in flat (10 or 26 dimensional) space-time. The results we will present here about strings in strong gravitational fields support this suggestion.

Since the most relevant new physics provided by strings concerns quantization of gravity, we must, at least, understand string quantization in a curved space-time. Actually, one would like to extract the space-time and the particle spectrum from the solution of string theory, but we are still far from doing that explicitly.

Practically, all what we know about strings comes from their study in flat critical (10 or 26) dimensional space-time. It must be noticed that expanding in perturbation around the Minkowski metric is not better since the non-trivial features appear in the strong curvature regimes, in the presence of horizons and of the intrinsic singularities. Curved space-times, besides their evident relevance in classical gravitation are also important at energies of the order of the Planck scale. At such energy scales, the picture of particles propagating in flat space-times is no longer valid and one must take into account the curved geometry created by the particles themselves. In other words, gravitational interactions are at least as important as the rest and can not anymore being neglected as it is usually the case in particle physics.

As a first step in the understanding of quantum gravitational phenomena in the framework of string theory, we started in 1987 a programme of string quantization in curved space-times. A summary of the developpements and results till now in this programme is given in what follows.

STRINGS IN PHYSICALLY RELEVANT CURVED SPACE-TIMES

Until now, gravity has not completely been incorporated in string theory: strings are more frequently formulated in *flat* space-time. Gravity appears through massless spin two-particles (graviton). One disposes only of partial results for strings in curved backgrounds, and these mainly concern the problem of consistency (validity of quantum conformal invariance) through the vanishing of the beta-functions. The non linear quantum string dynamics in curved space-time has only been studied in the slowly varying approximation for the geometry (background field method) where the field propagator is essentially taken as the flat-space Feynman propagator.
Clearly, such approximations are useless for the computation of physical quantities (*finite parts*) such as the mass operator, scattering amplitudes and critical dimension in strong-curvature regimes. Our aim is to properly understand strings in the context of gravity (classical and quantum). As a first step in this program we proposed to study Q.S.T. (Quantum String Theory) in curved space-times. There are different kind of effects to be considered here :
1) ground state and thermal effects : these are associated to the fact that in general relativity there are no preferred reference frames, and one has the possibility of having different choices of time. This arises the possibility for a given quantized field or string theory to have different alternative well defined Focks spaces (different "sectors" of the theory),

(which may be or may be not related by Hawking radiation). Associated with this: The presence of "intrinsic" statistical features (temperature, entropy) arising from the non-trivial structure (geometry, topology) of the space-time and not from a superimposed statistical description of the quantum matter fields themselves. We will not discuss these effects here. For a detailed account of them see refs [2-4].

2) Curvature effects : these will modify the mass spectrum, the critical dimension and the scattering amplitudes of the strings. We will discuss them below.

3) Conceptual aspects : In addition to those discussed above there are conceptual aspects to consider here, which appear when strings are restricted to live in causally disconnected regions of the space-time, that is in the presence of event horizons, and which imply quantum fluctuations of the event horizon and of the light-cone itself.

4) Finally, a word of warning on *the question of conformal invariance* in the metrics we consider. It is well known [5] that for a single string moving in a background, the conditions of quantum conformal invariance coincide with the *vacuum* Einstein equations (modulo string corrections). As such they require, at tree level, $R_{\mu\nu}$ =0. Certainly, de Sitter space time , cosmological backgrounds, gravitational shock-waves,... do not satisfy such a condition. Our interpretation of this point [6] is that such backgrounds are simply *not* candidate string ground states (vacua). Yet these physically relevant metrics can play the role of effective backgrounds felt by a single string as it moves in the presence of many others. An example of a situation of this kind is the effective Aichelburg-Sexl (AS) metric felt by one string as it collides with another at very high (Planck) energies [7-9]. Here too one would not identify the AS metric as a possible string vacuum; nonetheless, this metric is physically relevant to the description of the planckian energy collision process.

The price to pay for simplifying the true many-body problem into that of a single string moving in a effective, external metric will be indeed some (hopefully small) violation of unitarity. The main physics conclusions should, however, retain their validity at some semi-quantitative level, especially in the region of validity of a semi-classical approximation (since the conformal anomaly is an O($\hbar^2$) effect).

In flat space-time, the string equations of motion are linear and one can solve them explicitely, as well as the quadratic constraints. It should be recalled that the constraints in string theory contain as much physical information as the equations of motion. In curved space-time, the string equations of motion are highly non-linear (these equations are of the type of non-linear sigma models) and they are coupled to the constraints. Thus, right and left movers interact with each other and also with themselves. In flat Minkowski space-time, it is always possible to choose a gauge in which the physical time $X^0(\sigma,\tau)$ (or a light cone combination of it) and the world -sheet time τ are identified.   In curved space-time, the relation between the world sheet time τ and the physical time $X^0(\sigma,\tau)$ is involved, and in general not exactly known. However, it is possible to find the proportionality between τ and X^0 in some well defined asymptotic regimes or space-time regions. For all the physically relevant space-times we have studied, we have found well defined regimes in which $X^0(\sigma,\tau) \approx \tau$.

In ref [10], we have proposed a general scheme to solve the string equations of motion and constraints, both classically and quantum mechanically. The principle is the following: we start from an exact particular solution and develop in perturbations around it. We set

$$X^A(\sigma,\tau) = q^A(\sigma,\tau) + \eta^A(\sigma,\tau) + \zeta^A(\sigma,\tau) + \ldots$$

Here $q^A(\sigma,\tau)$ is an exact solution of the string equations and $\eta^A(\sigma,\tau)$ obeys the linearized perturbation around $q^A(\sigma,\tau)$. $\zeta(\sigma,\tau)$ is a solution of the second order perturbation around $q^A(\sigma,\tau)$. Higher order perturbations can be considered systematically. The choice of the starting solution is upon physical insight. Usually, we start from the solution describing the center of mass motion of the string $q^A(\tau)$, that is the point particle (geodesic) motion. The world-sheet time variable is here identified with the proper time of the center of mass trajectory. Even at the level of the zero order solution, gravitational effects including those of the singularities of the geometry are fully taken into account. It must be noticed that in our method, we are treating the space-time geometry *exactly* and taking the string oscillations around q^A as perturbation. So, our expansion corresponds to low energy excitations of the string as compared with the energy associated to the geometry. In a cosmological or in a black hole metric, our method corresponds to an expansion in ω/M, where ω is the string frequency mode and M is the universe mass or the black hole mass respectively. This can be equivalently considered as an expansion in powers of $(\alpha')^{1/2}$. Actually, since $\alpha' = (\ell_{Planck})^2$, the expansion parameter turns out to be the dimensionless constant

$$g = \ell_{Planck}/R_c = 1/(\ell_{Planck} M) = \omega/M$$

where R_c characterizes the curvature of the space-time under consideration and M its associated mass (the black-hole mass or the mass of the Universe in the cases before mentioned). So, our expansion is well suited to describe strings (test strings) in strong gravitational fields. In most of the interesting situations, one clearly has $g \ll 1$.

The constraint equations must also be expanded in perturbations. The classical (mass)^2 of the string is defined through the center of mass motion (or Hamilton-Jacobi) equation. The conformal generators (or world-sheet two dimensional energy-momentum tensor) are bilinear in the fields $\eta^A(\sigma,\tau)$. In order to obtain these constraints to the lowest non-trivial order it is necessary to keep first- and second-order fluctuations. Notice the difference with field theory where the first order fluctuations are enough to get the leading order approximation around a classical solution.

If we would like to apply our method to the case of flat spacetime, the zero order solution $q^A(\tau)$ plus the first order fluctuations $\eta^A(\sigma,\tau)$ would provide the exact solution of the string equations.

Strings in black-hole space times

We have applied our method to describe the quantum string dynamics in the Schwarzschild geometry and computed the effects of the scattering and interaction between the string and the black hole [11]. We have analyzed the string equations of motion and constraints both in the Schwarzschild and Kruskal manifolds and their asymptotic behaviours. The center of mass motion is explicitely solved by quadratures. We found the first and second order quantum fluctuations, η and ζ around the center of mass solution. We give an "in" and "out" formulation of this problem. We define an in-basis of solutions in which we expand first and second-order quantum fluctuations and define left and right oscillation modes of the string in the asymptotically flat regions of the space-time. The ingoing solutions are defined by selecting the behaviour at $\tau \to -\infty$ equal to a purely positive frequency factor $e^{-in\tau}$ (in-particle states). The ingoing ($\tau \to -\infty$) and outgoing ($\tau \to +\infty$) coefficient modes are related by a linear-Bogoliubov transformation describing transitions between the internal oscillatory modes of the string as a consequence of the scattering by the black hole. We find two main effects: (i) a change of *polarization* of the modes without changing their rigth or left character and (ii) a *mixing* of the particle and antiparticle modes changing at the same time their right or left character. That is, (i) if in the ingoing state, the string is in an excited mode with a given polarization, then in the outgoing state, there will be non-zero amplitudes for modes polarized in any direction and (ii) an amplitude for an *antiparticle* mode polarized in the same direction but with the right (or left) character reversed.

We have studied the conformal generators (L_n) and the constraints. An easy way to deal with the gauge invariance associated with the conformal invariance on the world sheet is to take the light-cone gauge in the ingoing region. Time evolution from $\tau \to -\infty$ to $\tau \to +\infty$ conserves the physical or gauge character of the modes. The independent physical excitations are those associated with the transverse modes. We solve for the second-order fluctuations in a mode representation, and then we get the conformal generators. These generators can be computed in terms of the ingoing basis ($\tau \to -\infty$) or alternatively, with the outgoing basis ($\tau \to +\infty$). The conservation of the two-dimensional energy-momentum tensor yields $L_n^{in} = L_n^{out} + \Delta L_n$, where ΔL_n describes excitations between the internal (particle) states of the string due to the scattering by the black hole. We find the mass spectrum from the $L_0 \sim 0$ constraint, which is formally the *same* as in flat space-time. This is a consequence of the *asymptotically flat* character of the space-time and of the absence of bound states for D>4. (If bound states would exist, they would appear in the (mass)^2 operator besides the usual flat space spectrum). The critical dimension at which massless spin-two states appear is D=26, the *same* as in flat space-time.

We have studied the elastic and inelastic scattering of strings by a black hole. The Bogoliubov coefficient A_n describes elastic processes and B_n inelastic ones. By elastic amplitudes we mean that the initial and final states of the string corresponding to the nth mode are the same. We find that pair creation out of the in-vacuum takes place for $\tau \to +\infty$ as a consequence of the scattering by the black hole. (Each pair here is formed by a right and left mode). The explicit computation of the coefficients A_n and B_n has been performed in an expansion at first order in $(R_s/b)^{D-3}$, R_s being the Schwarzschild radius and, b the impact parameter of the center of mass of the string. The elastic scattering cross section of the string by the black hole is given by

$$\left(\frac{d\sigma}{d\Omega}\right)_{elastic} = \left(\frac{d\sigma}{d\Omega}\right)_{c.m} |< n^{out} | n^{in} >|^2 = \left(\frac{d\sigma}{d\Omega}\right)_{c.m} \left[1 + \sum_{n' \neq n} | B_{n'}|^2 \right]$$

$$\left(\frac{d\sigma}{d\Omega}\right)_{elastic} = \left(\frac{d\sigma}{d\Omega}\right)_{c.m} \left[1 + \left(\frac{\pi p \alpha'}{b}\right)^2 \left(\frac{R_s}{b}\right)^{2(D-3)} \right] \sum_{m' \neq n} \left| F\left(\frac{m}{p}, \frac{n'b}{\alpha' p}\right) \right|^2$$

where $\left(\frac{d\sigma}{d\Omega}\right)_{c.m} = C_D \; R_s^{D-2} \; \Theta^{-(D-1 + 1/(D-3))} \left[1 + \left(\frac{D-3}{D-2}\right)\frac{m^2}{p^2} \right]^{(D-2)/(D-3)}$

$$C_D = (D-3) \left[(\pi)^{1/2} \Gamma(D/2)/\Gamma((D-1)/2) \right]^{(D-2)/(D-3)}$$

Here, the function F is a dimensionless number [11] and α' is the string tension. The factor C_D is typical of D-dimensions. $(d\sigma/d\Omega)_{c.m}$ is the center of mass scattering cross section for large impact parameter b and small scattering angle Θ. For D=4, we recover the analogue of Rutherford's formula:

$$\Theta_{D=4} = \frac{R_s}{b}\left(2 + \frac{m^2}{p^2}\right) \; , \; \frac{d\sigma}{d\Omega}_{D=4} \underset{\Theta \ll 1}{=} \frac{4 R_s^2}{\Theta^4} \left(1 + \frac{m^2}{2p^2}\right)^2$$

We see that quantum string corrections to the Rurtherford's scattering are of order α'^2. Quantum corrections to the scattering of particles by a black hole have been obtained previously in the framework of point-like particles [12].

It should be noticed that in the point particle theory, the interaction of particles with the Schwarzschild geometry is static. Pair creation only takes place via Hawking radiation. In contrast, the interaction of strings with the Schwarzschild geometry exhibits *new* features due to the *composite* character of the strings. The infinite set of oscillator modes constituting the string becomes excited during the scattering by the influence of the black hole field. Actually, any localized external field would lead to qualitatively similar effects. As a result, particle transitions between the ingoing and the outgoing final states take place, giving rise to the phenomenom of *particle transmutation* [11,13]. A more detailed description of this phenomenon is given below. This effect is proportional to B_n^2 and describes inelastic scattering since the final state is different from the initial one. It must be noticed that B_n is of order of α' and therefore this is a very small effect at least energies of order of Planck energy are reached. This particle transmutation effect is a genuine *stringy* effect which *does not* exist in the context of point-particle field theory. On the other hand, this effect is *not* related to Hawking radiation and the presence of an event horizon is not essential here. The Hawking radiation and related phenomena appearing in usual quantum field theory have an analogue in string theory [2-4]. They are linked to the possibility for a given field or string theory to have different non-equivalent descriptions (different choices of the physical time and thus different possible definitions of particle and vacuum states).

Particle transmutation from the scattering of strings and superstrings in curved space-time

More recently, we have given a general formulation of the scattering of strings by a curved space-time, for both open and closed bosonic strings and for supersymmetric ones [13]. We consider space-times which admit flat regions in order to define ingoing and outgoing scattering states, as it is the case for the shock-wave space-times, for the space-time of very thin and straight cosmic strings, and for asymptotically flat geometries. Due to the interaction with the geometry, the string excitation state changes from the ingoing to the outgoing situation. Therefore, if the string ingoing state described a given particle with mass m and spin s, the outgoing state may describe a different particle with mass m' and spin s'. When (m,s) = (m',s') this is an elastic process, and we have the same particle state in the initial and final states, although the momentum and spin polarization may change. Otherwise, the process is *inelastic;* the initial particle (m,s) *transmutes* into a different final one (m',s'). At first order in $(\alpha')^{1/2}$, (α' being the string tension), for string oscillations small compared with the energy scales of the metric, outgoing and ingoing oscillator operators are related by a linear, or Bogoliubov transformation. For open strings, transitions from the ground state to a state with an *even* number of creation operators are non-zero. (These final states describe very heavy particles of mass greater than the Planck mass). Transitions from the ground state to states with an *odd* number of creation operators *vanish*. For closed strings, the more relevant transitions are those from the ground state to a dilaton, to a graviton and to massive states, and the transmutation of a *dilaton into a graviton.* For supersymmetric strings, massless particles cannot transmute among themselves (this is true to all orders in $(\alpha')^{1/2}$). Several properties of the particle transmutation processes can be derived directly from the symmetry properties of the geometry. Particle transmutation amplitutes for strings in black hole spacetimes and gravitational shock-waves have been explicited computed in [13].

String propagation through gravitational shock waves

Recently, gravitational shock wave backgrounds have raised interest in the context of both field theory and strings [7-9,14-21]. These metrics are relevant to the particle scattering at the Planck energy scale. They correspond to boosted geometries in the limit in which the velocity of the source tends to the speed of light and the mass of the source tends to zero in an appropriate way. The ultrarelativistic limit of the Schwarschild solution is the Aichelburg-Sexl geometry (point particle source). The gravitational shock waves corresponding to the ultrarelativistic limit of the Kerr-Newmann geometry, to ultrarelativistic cosmic strings and other ultrarelativistic extended sources, have been recently found [17,18].
Remarkable enough, the string equations of motion [8,9] and the Klein-Gordon equation [14,8] have been *exactly* solved in this geometry. The mass spectrum and the critical dimension are the same as in flat space time but there is non-trivial elastic and inelastic scattering of the string by the shock wave [8]. We have found the *exact non-linear* transformation relating the ingoing ($\tau < 0$) and outgoing ($\tau > 0$) string mode operators (and zero modes) before and after the collision with the gravitational the shock wave. This transformation contains all the information about the scattering and interaction of the string with the shock wave geometry. The linearized transformation at first order in $(\alpha')^{1/2}$ is a Bogoliubov transformation. As in the black-hole case, transitions take place between the internal modes of the string. Here too, corrections to the point particle scattering cross section are of order α'^2. For large impact parameters, the scattering angle and cross section in the black hole and in the shock wave geometry are very similar.

More recently, we have performed this treatment for a general shock wave space time of *any localized source.* We have computed the *exact* expectation values of the total number and mass square operators of the string and show that they are *finite,* which generalize ours previous results in the Aichelburg-Sexl geometry [20,21]. We have studied the energy-momentum tensor of the string, computed the expectation values of all its components and show they are *finite.* The ingoing-outgoing ground state transition amplitude $< O_< | O_> >$ expresses as a sum of terms, which can be interpreted [20,21] as a n-leg scalar amplitude with vertex operators inserted at $\tau=0$ (a line of pinchs at the intersection of the world sheet with the shock wave), and for which we have found integral representations. The integrands posses equally spaced real pole singularities *typical* of string models in flat space-time. The presence and structure of these poles *is not* at all related to the structure of the space time geometry (which may or may be not singular). We give a sense to these integrals by taking the principal value prescription, yielding a well defined finite result. For the expectation values of the mass and number operators, we find similar integral representations. The integrands factorizes into two pieces: a part (given by the Fourier transform of the density matter of the source) which characterizes the shock wave geometry and the function

$$ \text{tg}(\alpha' \pi p^2) \, \Gamma(\alpha' p^2) / \Gamma(\alpha' p^2 + 1/2) \, , $$

which depends only of the string. This integrand posses real singularities (poles) like the tree level *string* spectrum. A *physical* interpretation of such poles is that they correspond to all higher string states which become excited after the collision through the shock wave. Moreover, the quantum expectation values of the string energy-momentum tensor $T^{AB}(X)$ admit similar integral representations (with the same structure of integrands and poles). Instead of studying the local dependence (as it would be appropriated for classical or cosmic strings), we integrate $T^{AB}(\vec{X}, X^0)$ over an spatial volume completely closing the string at time X^0, since the fundamental string here describes particle states. That is, we define

$$ \tau^{AB}(X^0) = \int (-G)^{1/2} \, T^{AB}(\vec{X}, X^0) \, d^{D-1}\vec{X} $$

For asymptotic times $|X^0| \to \infty$, we find

$$ \tau^{AB} = p^A p^B / p^0 \, , $$

which is precisely the energy-momentum tensor of a point particle integrated over a spatial volume. Even for $X^0 \to \infty$, the τ^{AB} of the string is not trivial because of the constraints $P^{0\,2} = {X'}^2 + X^2$. We find vacuum polarization effects induced by the shock wave on the string oscillators. There is a stress in the longitudinal and transverse directions. We have also computed the *fluctuations*

$$ (\Delta \tau^{AB})^2 = < \tau^{AB\,2} > - < \tau^{AB} >^2 $$

These fluctuations are *finite* and *non-zero* even for the AB components where $<\tau^{AB}> = 0$. In particular, for the energy density, the expectation value is trivial (it coincides with the flat space time value equal to the mass) but exhibits *non-trivial* quantum fluctuations.

We want here to compare these string results with those known for point particle QFT in shock-wave backgrounds [22]. For point particle QFT, no vacuum polarization effects arise in these backgrounds since the ingoing (<) and outgoing(>) creation and annhilation operators do not get mixed (there is no Bogoliubov transformation in such context). Therefore, no particle creation effects takes place for point particle field theories in these geometries. On the contrary, for strings particle transmutations as well as polarization effects on the energy-momentum tensor appear in shock-wave space times. These effects can be traced back to the mixing of creation and annhilation in (<) and out (>) string oscillators.

Strings in Cosmological Backgrounds

We have quantized strings in de Sitter space-time first [10] . We have found the mass spectrum and vertex operator. The lower mass states are the same as in flat space-time up to corrections of order g^2 but heavy states *deviate* significantly from the linear Regge trajectories. We found that there exists a *maximum* (very large) value of order $1/g^2$ for the quantum number N and spin J of particles. There exists real mass solutions only for
$$N < N_{max} = \pi / 2g^2 + O(g^{-2/3}) \quad , \quad g = 10^{-61} .$$
Moreover, for states in the leading Regge trajectory, the mass monotonously increases with J up to the value
$$J_{max} = 1/g^2 + O(1)$$
corresponding to the maximal mass $m^2_{max} = 0.76 + O(g^2)$. Beyond J_{max} the mass becomes complex. These complex solutions correspond to unstable states already present here at the tree (zero handle) level.

From the analysis of the mass spectrum, we find that the critical dimension for bosonic strings in de Sitter space-time is D=25 (instead of the value 26 in Minkowski space-time). This result is confirmed by an independent calculation of the critical dimension from the path intergral Polyakov's formulation, using heat-kernel techniques: we find that the dilaton β- function in D-dimensional de Sitter space-time must be
$$\beta^\phi = (D + 1 - 26) / (\alpha' 48\pi^2) + O(1) .$$
It is a general feature of de Sitter space-time to lower the critical dimensions in one unit. For fermionic strings we find D=9 instead of the flat value D= 10.

We have found that for the first order amplitude $\eta^i(\sigma, \tau)$, (i = 1,....D-1 refers to the spatial components) , the oscillation frequency is
$$\omega_n = [n^2 - (\alpha' mH)^2]^{1/2} ,$$
instead of n , where H is the Hubble constant. For high modes $n \gg \alpha' mH$,

the frequencies $\omega_n \sim n$ are real. The string shrinks as the universe expands. This shrinking of the string cancels precisely the expanding exponential factor of the metric and the invariant spatial distance does not blow up. Quantum mechanically, these are states with real masses ($m^2 H^2 < 1$). This corresponds to an expansion time H^{-1} very much bigger than the string period $2\pi/n$, that is, many string oscillations take place in an expansion period H^{-1} (in only one oscillation the string does not see the expansion).

For low modes $n < \alpha' mH$, the frequencies become imaginary. This corresponds to an expansion time very short with respect to the oscillation time $2\pi/n$ ("sudden" expansion, that is the string "does not have time" to oscillate in one time H^{-1}). These *unstable* modes are analyzed as follows. The n=0 mode describes just small deformations of the center of mass motion and it is therefore a physically irrelevant solution. When $\alpha' mH > 1$ relevant unstable modes appear. Then, the n=1 mode dominates $\eta^i(\sigma,\tau)$ for large τ. Hence, if $\alpha' mH > (2)^{1/2}$, η^i diverges for large τ, that is fluctuations become larger than the zero order and the expansion breaks down. However, the presence of the above instability is a true feature as it has been confirmed later by further analysis [6].

The physical meaning of this instability is that the string grows driven by the inflationary expansion of the universe. That is, the string modes couples with the universe expansion in such a way that the string inflates together with the universe itself. This happens for inflationary (ie accelerated expanding) backgrounds. In ref. [6] we have studied the string propagation in Friedman-Robertson-Walker (FRW) backgrounds (in radiation as well as matter dominated regimes) and interpreted the instability above discussed as Jeans-like *instabilities*. We have also determined under which conditions the universe expands, when distances are measured by stringy rods. It is convenient to introduce the *proper amplitude* $\chi^i = C \eta^i$, where C is the expansion factor of the metric. Then, χ^i satisfies the equation

$$\ddot{\chi}^i + [n^2 - \ddot{C}/C] \chi^i = 0$$

Here dot means τ-derivative. Obviously, any particular (non-zero) mode oscillates in time as long as $\ddot{C}/C$ remains < 1 and, in particular, when $\ddot{C}/C < 0$. A time-independent amplitude for χ is obviously equivalent to a fixed proper (invariant) size of the string. In this case, the behaviour of strings is stable and the amplitudes η shrink (like $1/C$).

It must be noticed that the time component, χ^0 or η^0, is always well behaved and no possibility of instability arises for it. That is the string time is well defined in these backgrounds.

i) For non-accelerated expansions (e.g. for radiation or matter dominated FRW cosmologies) or for the high modes $n \gg \alpha' Hm$ in de Sitter cosmology, string instabilities do not develop (the frequencies $\omega_n \sim n$ are real). Strings behave very much like point particles: the centre of mass of the string follows a geodesic path, the harmonic-oscillator amplitudes η shrink as the univers expands in such a way to keep the string's proper size constant. As expected, the distance between two strings increase with time, relative to its own size, just like the metric scale factor C.

ii) For inflationary metrics (e.g. de Sitter with large enough Hubble constant), the proper size of the strings grows (like the scale factor C) while the co-moving amplitude η remains fixed ("frozen") , i.e. η ~ η (σ) .

Although the methods of references [10] and [6] allow to detect the onset of instabilities, they are not adequate for a quantitative description of the high instable (and non-linear) regime. In ref. [23] we have developped a new quantitave and systematic description of the high instable regime. We have been able to construct a solution to both the non-linear equations of motion and the constraints in the form of a systematic asymptotic expansion in the large C limit, and to classify the (spatially flat) Friedman-Robertson Walker (FRW) geometries according to their compatibility with stable and/or unstable string behaviour. An interesting feature of our solution is that it implies an asymptotic proportionality between the world sheet time τ and the *conformal time* T of the background manifold. This is to be contrasted with the stable (point-like) regime which is characterized by a proportionality between τ and the *cosmic time*. Indeed, the conformal time (or τ) will be the small expansion parameter of the solution: the asymptotic regime (small τ limit) thus corresponds to the large C limit only if the background geometry is of the inflationary type. The non linear, high unstable regime is characterized by string configurations such that

$$|X'^0| \ll |\dot{X}^0| \quad , \quad |\dot{X}^i| \ll |X'^i|$$

with
$$X^0(\sigma, \tau) = C L(\sigma) \quad , \quad L(\sigma) = (\delta_{ij} X'^i X'^j)^{1/2}$$

$$X^i(\sigma,\tau) = A^i(\sigma) + \tau^2 D^i(\sigma)/2 + \tau^{1+2\alpha} F^i(\sigma)$$

where A^i, D^i and F^i are functions determined completely by the constraints, and α is the time exponent of the scale factor of the metric : $C = \tau L^{-\alpha}$.

For power-law inflation : $1 < \alpha < \infty$, $X^0 = \tau L^{1-\alpha}$. For de Sitter inflation : $\alpha = 1$, $X^0 = \ln(-\tau HL)$. For Super-inflation: $0 < \alpha < 1$, $X^0 = \tau L^{1-\alpha}$ + const. Asymptotically, for large radius C → ∞, this solution describes string configurations with expanding proper amplitude.

These highly unstable strings contribute with a term of negative pressure to the energy-momentum tensor of the strings. The energy momentum tensor of these highly unstable strings (in a perfect fluid approximation) yields to the state equation $\rho = -P(D-1)$, ρ being the energy density and P the pressure (P < 0). This description corresponds to *large radius* C → ∞ of the universe.

For *small radius of the universe*, highly unstable string configurations are characterized by the properties

$$|\dot{X}^0| \gg |X'^0| \quad , \quad |X'^i| \ll |\dot{X}^i|$$

The solution for X^i admits an expansion in τ similar to that of the large radius regime. The solution for X^0 is given by L/C, which corresponds to *small radius* $C \to 0$, and thus to small τ. This solution describes, in this limit, string configurations with shrinking proper amplitude, for which CX'^i behaves asymptotically like C, while CX^i behaves like C^{-1}. Moreover, for an ideal gas of these string configurations, we found:
$\rho = P(D-1)$, with *positive pressure* which is just the equation of state for a gas of massless particles.

More recently [24], these solutions have been applied to the problem in which strings became a dominant source of gravity. In other words, we have searched for solutions of the Einstein plus string equations. We have shown, that an ideal gas of fundamental strings is not able to sustain, alone, a phase of isotropic inflation. Fundamental strings can sustain, instead, a phase of anisotropic inflation, in which four dimensions inflate and, simultaneously, the remaining extra (internal) dimensions contract. Thus, fundamental strings can sustain, simultaneously, inflation and dimensional reduction. In ref. [24] we derived the conditions to be met for the existence of such a solution to the Einstein and string equations, and discussed the possibility of a successful resolution of the standard cosmological problems in the context of this model.

<u>Strings falling into space-time singularities and gravitational plane-wave backgrounds</u>

Recently [25], we have studied strings propagating in gravitational-plane wave space-times described by the metric

$$dS^2 = F(U, X, Y) dU^2 - dU\, dV + dX^i\, dX^j ,$$

where $F(U, X, Y) = W(U)(X^2 - Y^2)$ and $W(U \to 0) = \alpha / U^\beta$.

U and V are ligth cone variables; α and β are positive constants. These are vacuum space-times. The space time is singular on the null plane U=0.

The string equations in this class of backgrounds are linear and *exactly solvable*. In the light cone gauge $U = \alpha' p \tau$ and after Fourier expansion in the world sheet coordinate σ, the Fourier components $X_n(\tau)$ and $Y_n(\tau)$ satisfy a one-dimensional Schrodinger-type equation but with τ playing the role of the spatial coordinate and $p^2 W(\alpha' p \tau)$ as the potential [26]. (Here p stands for the U-component of the string momentum). We studied the propagation of the string when it approaches the singularity at U=0 from U< 0. We find different behaviours depending on whether $\beta < 2$ or $\beta \geq 2$. For $\beta<2$, the string coordinates X and Y are regular everywhere, that is, the string propagates smoothly through the gravitational singularity U=0. For strong enough singularity ($\beta \geq 2$), the string goes off to X=∞ grazing the singularity plane U=0. This means that the string *does not* go accross the gravitational wave, that is the string *can not reach* the U>0 region. For particular initial configurations, the string remains *trapped* at the point X=Y=0 in the gravitational wave singularity U=0. The case in which $\beta=2$ and then $W(U) = \alpha/U^2$ is explicitely solved in terms of Bessel functions.

The string propagation in these singular space-times has common features with the fall of a point particle into a singular attractive potential $-\alpha/x^\beta$. In both cases, the falling takes place when $\beta \geq 2$. The behaviour in τ of the string coordinates $X^A(\sigma,\tau)$ is analogous to the behaviour of the Schrödinger equation wave function $\Psi(x)$ of a point particle. However, the physical content is different. The string coordinates $X(\sigma,\tau)$ are dynamical variables and not wave functions. Moreover, our analysis also holds for the quantum propagation of the string: the behaviour in τ is the same as in the classical evolution with the coefficients being quantum operators. At the *classical*, as well as at the *quantum* level, the string propagates or does not propagate through the gravitational wave depending on whether $\beta < 2$ or $\beta \geq 2$, respectively. In other words, *tunnel effect does not takes* place in this string problem.

It must be noticed that for $\tau \to 0-$, i.e. $U \to 0-$, the behaviour of the string solutions is *non-oscillatory in* τ whereas for $\tau \to \infty$, the string oscillates. This *new type* of behaviour in τ is analogous to that found recently for strings in cosmological inflationary backgrounds [23, 24]. For $\beta = 2$, it is possible to express the coefficients characterizing the solution for $\tau \to 0$ in terms of the oscillator operators for $\tau \to \infty$.

It must be noticed that the spatial (i.e. fixed $U = \tau$) proper length of the string grows indefinitely for $\tau \to 0$ when the string approaches the singularity plane. Here too, this phenomenon is analogous to that found for strings in cosmological inflationary backgrounds. Moreover, this analogy can be stressed by introducing for the plane-wave background the light-cone coordinate U by

$$dU = W(U) dU \quad , \quad \text{i.e.} \quad U = (\alpha)^{1/2} |U|^{1-\beta/2} / (1-\beta/2)$$

Then, U is like the conformal time in cosmological backgrounds. For instance, $W(U) = \alpha/U^2$ mimics de Sitter space for $\tau \to 0$, and in this case we have, as in de Sitter space, $U=(\alpha)^{1/2} \ln (\alpha p \tau)$.

We label with the indices < and > the operators in the regions U < T and $U > T$ respectively (i.e. before and after the collision with the singularity plane U=0). We compute the total mass squared and the total number of modes $< N_> >$ after the string propagates through the singularity plane U=0 and reaches the flat spacetime region $U > T$. This has a meaning only for $\beta < 2$. For $\beta \geq 2$, the string does not reach the $U > T$ region and hence there are no operators $>$. In particular, there are no mass squared $M_>^2$ and total number $N_>$ operators for $\beta \geq 2$. For $\beta < 2$, $< M^2_> >$ and $< N_> >$ are given by [25]

$$< M_>^2 > = m_0^2 + 2\alpha'^{-1} \sum_{n=1}^{\infty} n \left[|B_n^x|^2 + |B_n^y|^2 \right]$$

$$< N_> > = 2 \sum_{n=1}^{\infty} \left[|B_n^x|^2 + |B_n^y|^2 \right]$$

where $B_n^x = -B_n^y \underset{n \to \infty}{\approx} (2n/\alpha' p)^{\beta-2}$, $0 < \beta < 1$

Here the expectation values refer to the ingoing (e.g., $<$) ground state $< M_>^2 >$ is finite for $\beta < 1$ but diverges for $1 \leq \beta < 2$. $< N_> >$ is finite for $\beta < 3/2$ but diverges for $3/2 \leq \beta < 2$. The physical meaning of these infinities is the following: These divergences are due to the infinite transverse extent of the wave-front and not to the short distance singularity of W(U) at U=0. The gravitational forces in the transverse directions (X,Y) transfer to the string a finite amount of energy when the transverse size of the shock wave front is finite. When the size of the wave front is infinite, the energy transfered by the shock-wave to the string impart large elongation amplitudes in the transverse directions (X,Y) which are responsible of the divergence of $< M_>^2 >$. This question has been analyzed in detail in ref.[24].

For a sourceless shock-wave with metric function

$$F(U, X, Y) = \alpha \, \delta(U) \, (X^2 - Y^2),$$

we find that $B_n^x = B_n^y = \alpha p \alpha' / 2in$ and these are the same coefficients as those corresponding to $W(U) = \alpha / U^\beta$ with $\beta = 1$. This is related to the fact that both functions W(U) have the same scaling dimension. The string propagation is formally like that of a Schrödinger equation with a Dirac delta potential: the string *passes across* the singularity at U=0 and *tunnel effect* is present. The string scattering in this sourceless shock wave is very similar to the string scattering by a shock-wave with a non-zero source density [20,21].

We have also computed $< M_>^2 >$ and $< N_> >$ for a metric function

$$F(U, X, Y) = \alpha \, \delta(U) \, (X^2 - Y^2) \, \theta(\rho_0 - X^2 + Y^2)$$

where θ is the step function and ρ_0 gives the transverse size of the wave-front. This F belongs to the shock-wave class with a density source we have treated in refs. [20,21]. Here too $<M_>^2>$ is finite as long as ρ_0 is finite. This shows explicitly that the divergence of $<M_>^2>$ is due to the infinite transverse extent of the wave-front and not to the short distance singularity of W(U) at U=0. More generally, for a string propagating in a schock wave spacetime with generic profile

$$F(U, X, Y) = \delta(U) f(X, Y),$$

we have found the exact expressions of $<M_>^2>$ and $<N_>>$ in ref. [20,21]. When f(X,Y) has infinite range, the gravitational forces in the X,Y directions have the possibility to transfer an infinite amount of energy to the string modes. We have computed all the components of the string energy-momentum tensor near the U=0 singularity.

The propagation of classical and quantum strings through these singular space-times is physically *meaningful* and provides new insigths about the physics of strings on curved space-times.

Strings in gravitational plane-wave backgrounds have been studied in ref.[26]. However, this problem has subtle points which were overlooked there. The analysis done in ref.[26] by analogy with the Schrödinger equation is not enough careful. The mass and number operators are expressed in terms of the transmission coefficient B_n. In ref. [26] the cases in which $B_n = \infty$, mean that there is no transmission to the region U>0, and then, there is no mass operator, neither number operator (since there is no string) in that region. This is the situation of falling to U=0 for $\beta \geq 2$ which we mentioned above. Therefore, $M_>^2$ and $N_>$ make sense only for $\beta < 2$ and any statement about $M_>^2$ and $N_>$ for $\beta \geq 2$ is meaningless.

Strings in topologically non-trivial backgrounds: scattering of a quantum (fundamental) string by a cosmic string

In ref [27], we have studied a (quantum) fundamental string in a conical space-time in D dimensions. This geometry describes a straight cosmic string of zero thickness and it is a good approximation for very thin cosmic strings with large curvature radius. The space-time is locally flat but globally it has a non-trivial (multiply connected) topology. There exists a conelike singularity with azimuthal deficit angle

$$\delta\Phi = 2\pi(1-\alpha) = 8\pi G\mu.$$

$G\mu$ is the dimensionless cosmic-string parameter, G is the Newton constant and μ the cosmic-string tension (mass per unit of length). $G\mu$ 10^{-6} for standard cosmic strings of grand unified theories.

The string equations of motion and constraints are exactly solvable in this background. The string equations are free equations in the Cartesian-type coordinates X^0, X, Y, Z^i ($3 \leq i \leq D-1$), but with the requirement that

$$0 \leq \arctan(Y/X) \leq 2\pi\alpha.$$

The exact solution in the light-cone gauge is given in ref.[27] .The string as a whole is deflected by an angle

$$\Delta = \delta\Phi / 2.$$

A string passing to the right (left) of the topological defect is deflected by + (-)$\delta\Phi$. This deflection does not depend on the impact parameter, nor on the particle energy due to the fact that the interaction with the space-time is of purely topological nature. In the description of this interaction we find essentially two different situations:
(i) The string does not touch the scatterer body. A deflection $+ \Delta$ at the origin and a rotation in the polarization of modes takes place. In this case there is no creation or excitation of modes (creation and annhilation operators are not mixed) and we refer to this situation as elastic scattering. (ii) The string collides against the scattering center; then in addition to being deflected, the internal modes of the string become excited. We refer to this situation as inelastic scattering. In the evolution of the system, continuity of the string coordinates and its τ-derivatives at the collision time $\tau = \tau_0$ is required. In this case, the relation between the ingoing and outgoing oscillators is given by an exact Bogoliubov transformation. In addition to a change in the polarization, there are mode excitations which yield final particle states *different* from the initial one. This provides another example of the particle transmutation process described above. Notice that we are dealing with a single (test) string. That is, the initial and final states are *one particle* but *different* states. Also notice that the particle states *transmute* at the classical (tree) level as a consequence of the interaction with the space-time geometry. In the present case, this is a topological defect.

We explicitly proved [27], that the conformal L_n generators $Ln_<$ built from the ingoing modes are identical to the $L_{n>}$ built from the outgoing modes. The mass spectrum is the same as in the standard Minkowski space-time and the critical dimension is the same (D=26 for bosonic strings).

Let us notice that strings in conical space-times were considered in ref. [28] but only for deficit angles $2\pi(1-1/N)$, N being an integer, where the scattering is *trivial*. (In that case, the space becomes an orbifold). In our work, we have solved the scattering problem for general deficit angles where it is nontrivial. Let us also notice that the condition of conformal invariance (vanishing of the β function) is identically satisfied everywhere in the conical space-time, except eventually at the origin. If such difficulty arises, this space-time will simply not be a candidate for a string ground state (vacuum). Anyway, this geometry effectively describes the space-time around a cosmic string.

Can the string split ? An interesting question here is whether the string may split into two pieces as a consequence of the collision with the conical singularity. When the string collides against the conical singularity, since the deflection angles to the right and to the left of the scattering center are different, one could think that the splitting of the string into two pieces will be favored by the motion. Such splitting solution exists and is consistent. However, its classical action is larger than the one without splitting. Therefore, this splitting may take place *only quantum mechanically*. In fact, such a possibility of string splitting always exists and already in the simplest case for strings *freely* propagating in *flat* space-time. The free equations of motion of strings in flat space-time admit consistent solutions which describe splitting but once more their action is larger than the solution without splitting.
When the string propagates in curved space-time the interaction with the geometry modifies the action. In particular, the possibility arises that the action for the splitting solution becomes smaller than the one without splitting. Therefore, string splitting will occur *classically*.

Finally, we have computed exactly and in closed form the scalar particle (lowest string mode) quantum scattering amplitude in the conical space of the cosmic string. For that, it is neccessary to know first, the solution of the Klein-Gordon equation in conical space-time in D dimensions. We have found the ingoing solution which satisfies the massive free equation with the non-trivial requirement to be periodic in the azimuthal angle Φ with period $2\pi\alpha$. This prevents the usual asymptotic behaviour for large radial coordinate $R \to \infty$. The full wave function is the sum of two terms. For D=3, this solution has been found in refs.[29] and [30]. The incident wave turns out to be a finite superposition of plane waves without distortion. They propagate following wave vectors rotated from the original one by a deflection $+\Delta$ and periodically extended with period $2\pi\Delta$. This incident wave although undistorted suffers multiple periodic rotations as a consequence of the multiply connected topology. In addition, the second term describes the scattered wave with scattering amplitude

$$f(\Theta) = \frac{1}{2\pi} \frac{\sin \pi/\alpha}{(\cos \pi/\alpha + \cos \Theta/\alpha)}$$

In the scalar (ground state) string amplitude, we have ingoing ($\tau < \tau_0$) and outgoing ($\tau > \tau_0$) zero modes and oscillator modes averaged in the ingoing ground state $|0_<>$. Inserting the in and out wave function solutions in this matrix element yields four terms, each of these terms splitting into four other ones corresponding to the natural four integration regions of the double τ-domain. The detailed computation is given in ref.[27]. (For this computation it is convenient to work in the covariant formalism where all string components are quantized on equal footing). The effect of the topological defect in space-time on the string scattering amplitudes manifests through the nontrivial vertex operator (which is different from the trivial one $e^{ik \cdot x}$, and through the fact that ingoing and outgoing mode operators are related by a Bogoliubov transformation which makes the expectation value on the ingoing ground state $|0_<\rangle$ non-trivial. In the $\alpha = 1$ limit (that is, for the cosmic string mass $\mu = 0$), we recover the flat space Minkowski amplitude. If the oscillator modes n=0 are ignored, we recover the point-particle field-theory Klein-Gordon amplitude.

REFERENCES

1. See for example: K.G. Wilson, Revs. Mod. Phys. 47, 773 (1975) and 55, 583 (1983).
2. N. Sánchez, Phys. Lett. B195, 160 (1987).
3. H.J. de Vega and N. Sánchez, Nucl. Phys. B299, 818 (1988).
4. N. Sánchez, "Quantum strings in curved space-times", Erice lectures May 2-12, 1989, in "Quantum Mechanics in curved space-time", J.Audretsch and V. de Sabbata, eds., Plenum Publ. Co., pp 265-315 (1991).
5. C. Lovelace, Phys. Lett. B135, 75 (1984);
   E.S. Fradkin and A.A. Tseytlin, Phys. Lett. B158, 316 (1985);
   C.G. Callan et al., Nucl. Phys. B262, 593 (1985).
6. N. Sánchez and G. Veneziano, Nucl. Phys. B333, 253 (1990).
7. D. Amati, M. Ciafaloni, and G. Veneziano, Int. J. of Mod. Phys. 7,1615 (1988).
8. H.J. de Vega and N. Sánchez, Nucl.Phys. B317, 706 and 731 (1989).
9. D. Amati and C. Klimcik, Phys.Lett. B210, 92 (1988).
10. H.J. de Vega and N. Sánchez, Phys. Lett. B197, 320 (1987).
11. H.J. de Vega and N. Sánchez, Nucl. Phys. B309, 552 and 577 (1988).
12. N. Sánchez, Phys. Rev. D18, 1798 (1978).
13. H.J. deVega, M. Ramon-Medrano and N. Sánchez, Nucl. Phys. B351, 227 (1991).
14. G. t' Hooft, Phys. Lett. B198, 61 (1987).
15. G. Veneziano, Mod. Phys. Lett. A2, 899 (1987);
    V. Ferrari, P. Pendenza and G. Veneziano, Gen. Rel. Grav. 20, 1185 (1988).
16. C.O. Lousto and N. Sánchez, I.J.M.P A5, 915 (1990).
17. C.O. Lousto and N. Sánchez, Phys. Lett. B232, 462 (1989).
18. C.O. Lousto and N. Sánchez, Meudon-DEMIRM preprint 91007 (to appear in Nucl. Phys. B).
19. G.T. Horowitz and A.R. Steif, Phys. Rev. Lett. 64, 260 (1990).
20. H.J. de Vega and N. Sánchez, Phys. Rev. Lett. (C) 65, 1567 (1990).
21. H.J. de Vega and N. Sánchez, Phys. Lett. B244, 215 (1990) and LPTHE preprint 90-47.

22. G. Gibbons, Comm. Math. Phys .45, 191 (1975).
    S. Deser, J. Phys. A8, 1972 (1975).
23. M. Gasperini, N. Sánchez and G. Veneziano, CERN-TH 5893/90, DFTT-30/90 and Meudon-DEMIRM 90091 preprint (to appear in IJMPA).
24. M. Gasperini, N. Sánchez and G. Veneziano, CERN-TH 6010/91, DFTT 06/91 and Meudon-DEMIRM 91004 preprint.
25. H.J. de Vega and N. Sánchez, LPTHE 90-48 preprint.
26. G.T. Horowitz and A.R.Steif, Phys. Rev. D42, 1950 (1990).
27. H.J. de Vega and N. Sánchez, Phys. Rev. D42, 3969 (1990).
28. J. A. Bagger, C.G. Callan and J.A. Harvey, Nucl. Phys. B278, 550 (1986). See also L. Dixon, J. A. Harvey, C. Vafa and E. Witten, ibid. B261, 67 (1985); B274, 285 (1986);
    D. Gepner and E. Witten, ibid. B278, 493 (1986).
29. S. Deser and R. Jackiw, Commun. Math. Phys. 118, 495 (1988).
30. G.'t Hooft, Comm. Math. Phys. 117, 685 (1988).

PROJECTIVE UNIFIED FIELD THEORY IN CONTEXT WITH THE COSMOLOGICAL TERM AND THE VARIABILITY OF THE GRAVITATIONAL CONSTANT*

Ernst Schmutzer

Theoretical-Physics Institute
Friedrich Schiller University Jena
Jena, Germany

ABSTRACT

The 5-dimensional variant "Projective Unified Field Theory" leads to a constant gravitational coupling factor (gravitational constant) and a variable cosmological coupling factor ("cosmological constant"). This last fact could be interesting for quantum field theory in connection with the quantum fluctuations of the vacuum.

1. PHYSICAL BASIC EQUATIONS IN THE 5-DIMENSIONAL PROJECTIVE SPACE

At the conferences "GR9" (Jena 1980) and "Unified Field Theories of more than 4 Dimensions, including Exact Solutions" (Erice 1982) we repeated our argument for the Einstein-like 5-dimensional field equation [1,2,3]

$$R^{\mu\tau} - \frac{1}{2} g^{\mu\tau} \overset{5}{R} + D^{\mu\tau} = \kappa_0 \Theta^{\mu\tau} \qquad (1)$$

as the basis of our Projective Unified Field Theory (PUFT). This field equation has already been chosen and investigated by us – in contrast to Jordan [4] – since 1957 (Greek indices run from 1 to 5).

The quantities used have the following meaning:

$g^{\mu\tau}$ metric tensor, X^μ projective coordinates, $R^{\mu\tau}$ Ricci tensor, $\overset{5}{R}$ curvature invariant, $\Theta^{\mu\tau}$ energy projector of the non-geometrized matter (substrate), κ_0 Einstein's gravitational constant. Further

$$\text{a)} \quad D^{\mu\tau} = \lambda_0 S^C (g^{\mu\tau} + C s^\mu s^\tau) \quad \text{with} \quad \text{b)} \quad D^{\mu\tau}{}_{;\tau} = 0 \qquad (2)$$

is the 5-dimensional generalization of the cosmological term (λ_0 and C free constants, S amount of the 5-dimensional radius vector, $s^\mu = X^\mu/S$ radial unit vector).

---

* Dedicated to my friend Professor Dr. Peter Gabriel Bergmann on the occasion of his 75th birthday

Because of the well-known geometrical relation

$$(R^{\mu\tau} - \frac{1}{2}g^{\mu\tau}\overset{5}{R})_{;\tau} = 0 \qquad (3)$$

and (2b) from (1) results the 5-dimensional conservation law

$$\Theta^{\mu\tau}{}_{;\tau} = 0 . \qquad (4)$$

For the case of an electrically charged perfect fluid (5-dimensional perfect fluid) we succeeded in finding the energy projector

$$\Theta^{\mu\tau} = -e^{\sigma}(\mu U^{\mu}U^{\nu} + ph^{\mu\nu}) + (B - e^{\sigma}p)s^{\mu}s^{\nu}$$
$$- \frac{e_0 L_0}{cS_0^2}e^{-\frac{\sigma}{2}}\varrho_0(U^{\mu}s^{\nu} + U^{\nu}s^{\mu}) . \qquad (5)$$

Here are:
σ the scalaric field defined inversely by

$$S = S_0 e^{\sigma} \qquad (S_0 \text{ free constant}) , \qquad (6)$$

U^{μ} the 5-velocity with

$$U^{\mu}U_{\mu} = -c^2 \qquad (7)$$

and

$$h^{\mu\nu} = g^{\mu\nu} + \frac{1}{c^2}U^{\mu}U^{\nu} - s^{\mu}s^{\nu} \qquad (8)$$

the 5-dimensional metrical projection tensor.

Further are:
μ mechanical mass density, p pressure, ϱ_0 electric charge density, e_0 electric elementary charge, B and L_0 free constants.

2. PHYSICAL BASIC EQUATIONS IN THE 4-DIMENSIONAL SPACE-TIME

By projection of (1) from the projective space into space-time the following basic equations in Gauss units result (Latin indices run from 1 to 4):

Generalized gravitational field equation

$$\overset{4}{R}_{mn} - \frac{1}{2}g_{mn}\overset{4}{R} + D_{mn} = \kappa_0(E_{mn} + \Sigma_{mn} + \Theta_{mn}) , \qquad (9)$$

where

$$E_{mn} = \frac{1}{4\pi}(B_{mk}H^k{}_n + \frac{1}{4}g_{mn}B_{jk}H^{jk}) \qquad (10)$$

(B_{mk} electromagnetic field strength tensor, H_{kn} electromagnetic induction tensor) is the electromagnetic energy tensor (Minkowski tensor),

$$\Sigma_{mn} = -\frac{3}{2\kappa_0}(\sigma_{,m}\sigma_{,n} - \frac{1}{2}g_{mn}\sigma_{,k}\sigma^{,k}) \qquad (11)$$

the scalaric energy tensor and Θ_{mn} the substrate energy tensor. Furthermore the cosmolocial term D_{mn} with the cosmological coupling factor

$$\lambda = \lambda_0 S_0{}^C e^{(C-1)\sigma} \tag{12}$$

occurs.

Generalized electromagnetic field equations

$$H^{mn}{}_{;n} = \frac{4\pi}{c} j^m \quad \text{(inhomogeneous Maxwell system)}, \tag{13}$$

$$B_{<mn,k>} = 0 \quad \text{(cyclic Maxwell system)} \tag{14}$$

(j^m electric current density).
Furthermore the relation

$$\text{a)} \quad H_{mn} = \bar{\varepsilon} B_{mn} \quad \text{with} \quad \text{b)} \quad \bar{\varepsilon} = e^{3\sigma} \tag{15}$$

($\bar{\varepsilon}$ scalaric dielectricity resp. permittivity of the vacuum) holds.

Scalaric field equation

$$\sigma^{,k}{}_{;k} = -\frac{2}{3} \lambda_0 (C-1) S_0{}^C e^{(C-1)\sigma} + \kappa_0 (\frac{1}{8\pi} B_{kj} H^{kj} + \frac{2}{3} \vartheta). \tag{16}$$

The scalaric substrate density ϑ occurring here is given by

$$\vartheta = e^{-\sigma} \Theta^\mu{}_\mu - \frac{3}{2} \Theta^m{}_m. \tag{17}$$

Conservation laws

Further by projection from (4) the equation of motion

$$\Theta^{mn}{}_{;n} = -\frac{1}{c} B^m{}_n j^n + \vartheta \sigma^{,m} \quad \text{(energy-momentum conservation)} \tag{18}$$

and the electrical continuity equation

$$j^m{}_{;m} = 0 \quad \text{(electric charge conservation)} \tag{19}$$

result.

In the case of a 5-dimensional perfect fluid from (5) follows

$$\Theta^{mn} = -\mu u^m u^n - p(g^{mn} + \frac{1}{c^2} u^m u^n) \quad (u^m \text{ 4-velocity}). \tag{20}$$

Then (18) takes the form ($\{{}^m_{kn}\}$ Christoffel symbol)

$$\left(\mu + \frac{p}{c^2}\right)\left(u^m{}_{,n} + \{{}^m_{kn}\} u^k\right) u^n + \left[(\mu + \frac{p}{c^2}) u^n\right]_{;n} u^m$$
$$= \frac{1}{c} B^m{}_n j^n - (p^{,m} + \vartheta \sigma^{,m}) \tag{21}$$

with

$$\left[(\mu + \frac{p}{c^2}) u^n\right]_{;n} = \frac{1}{c^2}\left(\frac{dp}{d\tau} + \vartheta \frac{d\sigma}{d\tau}\right), \tag{22}$$

whereas (19) reads
$$(\varrho_0 u^m)_{;m} = 0 \,. \tag{23}$$

The transition to the limiting case of a point-like test particle leads from (21) to the equation of motion

$$m \left(\frac{du^m}{d\tau} + \left\{ \begin{matrix} m \\ kn \end{matrix} \right\} u^k u^n \right) = \frac{e}{c} B^m{}_n u^n - s_0 c^2 \left(\sigma^{,m} + \frac{u^m}{c^2} \frac{d\sigma}{d\tau} \right) \tag{24}$$

with

$$\text{a)} \quad m = \int \mu \, d^{(3)}V \,, \quad \text{b)} \quad e = \int \varrho_0 \, d^{(3)}V \,, \quad \text{c)} \quad s_0 = \frac{1}{c^2} \int \vartheta \, d^{(3)}V \tag{25}$$

(m mechanical mass; e electric charge; s_0 scalaric mass, abbreviated: scalmass).
For different reasons it is useful to introduce the (dimensionless) scalaric density parameter

$$\gamma = \frac{\vartheta}{\mu c^2} \tag{26}$$

and the (dimensionless) scalaric particle parameter

$$\Gamma = \frac{s_0}{m} \,. \tag{27}$$

In a detailed paper[5] we presented the basic equations of this section 2 in first order approximation.

3. PHYSICAL OUTCOME AND ITS INTERPRETATION

From the physical point of view the main results and predictions of PUFT are[2,3]:

- In contrast to other variants of 5-dimensional field theories (projective-relativistic theories or Kaluza-Klein type theories) this theory has been constructed such that the gravitational constant remains a true fundamental constant of nature.

- This theory predicts the appearance of scalaric dielectricity (polarization) $\bar{\varepsilon} = \exp(3\sigma)$ of the vacuum (15b).

- In the case of vanishing substrate (pure geometrized theory) this theory is not identical with the so-called Einstein-Maxwell theory (simple superposition of the Einstein theory and the Maxwell theory). The reason is given by the occurrence of the scalaric energy tensor Σ_{mn} in (9).

- In the case of electrically neutral matter ($j^m = 0$, $B_{mn} = 0$) and vanishing scalaric substrate density ($\vartheta = 0$) from (16) $\sigma =$ const can be concluded. This means that only in this special case the Einstein theory is contained in our theory.

- The validity of the cyclic (homogeneous) Maxwell system (14) implies the non-existence of magnetic monopoles as physical (non-singularity) sources in the usual physical sense.

- An important physical advantage of this theory results from the positive definiteness of the scalaric energy density $\Sigma_4{}^4$ for the static and stationary case of the scalaric field, i. e. antigravitational effects can only be expected for non-stationary fields.

- From (24) one immediately recognizes besides the electromagnetic Lorentz force the occurrence of a new scalaric force term.

- Investigations within the framework of the point-particle Lagrangian formalism as well as spinor theoretical approaches lead to the scalarically modified mass $m_s = m_0 \exp(-\frac{\sigma}{2})$ of a point-particle with the 5-dimensionally based rest mass m_0.

- The application of this theory to a homogeneous and isotropic cosmological dust model [6] leads – in contrast to the well-known Friedman model – because of the relation $\mu K^3 \sim \exp(\gamma\sigma)$ to the non-conservation of the mechanical matter (K world radius).

Apart from the experimental verfication or falsification of PUFT a pensive theoretical comment of fundamental importance should be made here:
The quantity scalmass s_0 of a particle (25c) which plays as a basically new (of course hypothetical) property of a particle a peculilar role is – in contrast to the electric charge e – no conservation quantity, since no corresponding continuity equation (conservation law) exists. This porperty of s_0 reminds us of the similar property of the mechanical mass m, also not being a conservation quantity.

4. COSMOLOGICAL TERM

As it is well known, in Einstein's gravitational theory the cosmological term reads

$$D_{mn}^{\text{Einstein}} = \lambda_c \, g_{mn} \qquad (28)$$

where $\lambda_c = \text{const}$ is the cosmological constant.

In our PUFT instead of this constant the variable cosmological coupling factor (12)

$$\lambda = \lambda_0 S_0^C e^{(C-1)\sigma}, \qquad (29)$$

dependent on the scalaric field σ, occurs. In the special case of choice $C = 1$ the Einstein term follows with

$$\lambda_c = \lambda_0 S_0 . \qquad (30)$$

In quasi-Newtonian approximation the scalaric field eqution (16) gives

$$\Delta\sigma = \frac{2}{3}\kappa_0\left[\vartheta + \frac{3}{8\pi}\left(\vec{H}\vec{B} - \vec{E}\vec{D}\right)\right] + C_0 + C_1\sigma, \qquad (31)$$

where

a) $C_0 = -\frac{2}{3}\lambda_0(C-1)S_0^C \gtreqless 0$, b) $C_1 = -\frac{2}{3}\lambda_0(C-1)^2 S_0^C \gtreqless 0$, (32)

i. e. C_0 and C_1 can be positive or negative.

In the case of spherical symmetry results from (31) for the exterior of a body (vacuum) the field equation

$$\frac{1}{r^2}\frac{d}{dr}\left(\frac{d\sigma}{dr}r^2\right) - C_1\sigma = C_0 \qquad (33)$$

(r radial coordinate) with the solution

$$\sigma = \frac{\alpha_0}{r}e^{(\overset{-}{+})\sqrt{C_1}r} - \frac{C_0}{C_1} \qquad (\alpha_0 = \text{const}) . \tag{34}$$

It seems to us that the variable cosmological coupling factor occurring automatically in our theory could be intersting in quantum field theory in connection with the vacuum fluctuations.

5. REFERENCES

1. Schmutzer, E.: Z. Physik <u>149</u> (1957) 329
2. Schmutzer, E.: Proceedings of the 9th International Conference on General Relativity and Gravitation (1980), Deutscher Verlag der Wissenschaften Berlin and Cambridge University Press Cambridge 1983 (ed. E. Schmutzer)
3. Schmutzer, E.: Unified Field Theories of more than 4 Dimensions Including Exact Solutions, World Scientific, Singapore 1983 (eds. V. DeSabbata and E. Schmutzer)
4. Jordan, P.: Göttinger Nachr. (1945) 74; Naturw. <u>33</u> (1946) 250; Ann. Physik (Leipzig) <u>1</u> (1947) 219
5. Schmutzer, E. and Kroll, P.: Ann. Physik (Leipzig) <u>47</u> (1990) 340
6. Schmutzer, E.: Class. Quantum Grav. <u>5</u> (1988) 353, 1215; Ann. Physik (Leipzig) <u>47</u> (1990) 219

THE INTRODUCTION OF THE COSMOLOGICAL CONSTANT

E.L. Schucking

Department of Physics
New York University
New York, N.Y. 10003

On February 4, 1917 Albert Einstein writes to Paul Ehrenfest (Kerszberg 1989, p155): "I have ... again perpetrated something about gravitation theory which somewhat exposes me to the danger of being confined to a madhouse." Four days later he revealed these maddening thoughts in the physico-mathematical division of the Royal Prussian Academy of the Sciences to his fellow academicians. A week later the Reichsdruckerei issued the Sitzungsbericht "Kosmologische Betrachtungen zur allgemeinen Relativitätstheorie" (Einstein 1917). Thus began modern cosmology.

The fourth section of Einstein's paper "Über ein an den Feldgleichungen der Gravitation anzubringendes Zusatzglied" (about an additional term to be fixed to the field equations of gravitation) introduced the cosmological constant λ. Instead of his field equation (13)

$$G_{\mu\nu} = -\kappa (T_{\mu\nu} - \frac{1}{2} g_{\mu\nu} T) \tag{1}$$

where $G_{\mu\nu}$ denotes the Ricci tensor he suggested now the equation (13a)

$$G_{\mu\nu} - \lambda g_{\mu\nu} = -\kappa (T_{\mu\nu} - \frac{1}{2} g_{\mu\nu} T) . \tag{2}$$

To motivate the introduction of this new constant of nature without a wisp of empirical evidence he wrote that his new extension was "completely analogous to the extension of the Poisson equation to

$$\Delta \phi - \lambda \phi = 4\pi K \rho \quad " . \tag{3}$$

This remark was the opening line in a bizarre comedy of errors.

Einstein's modified Poisson equation is now familiar to all physicists through the static version of Yukawa's meson theory which has the spherically symmetric vacuum solution

$$\phi = \frac{const}{r} e^{-r\sqrt{\lambda}} , \quad \lambda = (\frac{mc}{\hbar})^2 , \quad r = (x^2 + y^2 + z^2)^{1/2} . \tag{4}$$

Gravitation and Modern Cosmology
Edited by A. Zichichi *et al.*, Plenum Press, New York, 1991

But this equation had a deeper root. The Königsberg theoretician Carl Neumann (Neumann 1896) had proposed the modified Poisson equation to introduce an exponential cut-off for the gravitational potential. He thus anticipated Einstein's worry about the disastrous influence of distant stars on the potential. Einstein, apparently, was not aware of Neumann's work in 1917.

It is true that the Poisson equation modified by a term $-\lambda\phi$ (with a positive λ) on its left hand side leads to an exponential cut-off for the gravitational potential. But Einstein's flat assertion that the λ-term in his field equations had a completely analogous effect was wrong. However generations of physicists have parroted this nonsense. Even Abraham Pais (Pais 1982) writes in his magisterial Einstein biography about the analogy between the λ-terms in Poisson's and Einstein's equations "he (Einstein) performs the very same transition in general relativity".

It seemed so deceptively obvious: the potential corresponds in the Newtonian approximation to (c = 1)

$$g_{00} = -(1 + 2\phi).\tag{5}$$

Thus adding a term $-\lambda\phi$ to $\Delta\phi$ might correspond to inserting a term $-\lambda g_{\mu\nu}$ in addition to the Ricci tensor whose 00-component gives essentially the Laplacean in Newtonian approximation.

I still remember when Otto Heckmann told me 35 years ago: "Einstein's Argument ist naturlich Quatsch (baloney)". And the late Hamburg cosmologist was right. For ϕ is ϕ/c^2 and can be neglected compared to one in first approximation. Thus the Newtonian analog of Einstein's equations with λ-term is not the modified Poisson equation (3) but

$$\Delta\phi + \lambda c^2 = 4\pi K\rho.\tag{6}$$

With equation (6) Einstein had not introduced an exponential cut-off for the range of gravitation but a new repulsive force ($\lambda > 0$), proportional to mass, that pushed away every particle of mass m with a force

$$\vec{F} = mc^2 \frac{\lambda}{3}\vec{x},\tag{7}$$

a force derivable from the repulsive oscillator potential $-\lambda c^2 r^2/6$. This was clearly stated by Arthur Eddington (Eddington 1923). That follows simply from the equation of geodesic deviation.

Instead of getting a shielded gravitation one had now at large distances almost naked repulsion. This was quite different from the expected bargain.

I shall not relate the fascinating story of an actor named lambda in the ensuing scenes of the comedy. Pierre Kerszberg (Kerszberg 1989) in Sydney, Australia, has just published a 403 page book with the subtitle "The Einstein-De Sitter controversy (1916-17) and the Rise of Relativistic Cosmology" where you can learn what happened on stage and behind the scenes. The only justification for my note is that Kerszberg's admirable tome did not so bluntly expose the false pretense under which Herr Lambda insinuated himself into this potential play.

With keen hindsight we observe now that the introduction of lambda amounted to a redefinition of the vacuum state for the universe - the replacement of Minkowski space-time by de Sitter space-time.

Why had this new definition been introduced? When Einstein announced his theory of gravitation 75 years ago he thought he had generalized special relativity. He believed his theory showed that rotation was relative rotation against the distant masses and that inertia of a particle was an effect of distant matter, he thought he had exorcised absolute space, the God of Isaac Newton. He thought his paper did demonstrate all that.

It did no such thing. But it created the Einstein cosmos. His amended field equations allowed now a S^3 in which gravitation and cosmic repulsion equilibrated. This cosmos obtained as a solution of dynamical equations for the gravitational field created a new "System of the World". Modern cosmology deals with its variations.

ACKNOWLEDGMENTS

I am grateful to Jia-Zhu Wang for ingenious help.

REFERENCES

Eddington, A., 1923, "The Mathematical Theory of Relativity." Cambridge University Press. p161.

Einstein, A., 1917, Sitzungsberichte der K. Akademie VI p142.

Kerszberg, P., 1989, "The Invented Universe." Clarendon Press, Oxford.

Neumann, C., 1896, "Allgemeine Untersuchungen über das Newtonsche Prinzip der Fernwirkungen." Leipzig. p1.

Pais, A., 1982, "Subtle is the Lord ...". Clarendon Press Oxford. p286.

PRE-POST-HISTORY OF TOLMAN'S COSMOS

E.L. Schucking, S. Lauro[*], and J.-Z. Wang

Department of Physics, New York University, New York, N.Y. 10003
*Department of Physics, St. John's University, Jamaica, New York N.Y. 11451

In view of the objections raised by some reviewers concerning the use of the word "creation", it should be explained that the author understands this term, not in the sense of "making something out of nothing", but rather as "making something shapely out of shapelessness", as, for example, in the phrase "the latest creation of Parisian fashion".
  - Gamow, G., 1961, "The Creation of the Universe."

ABSTRACT

A unique extension of Tolman's radiation cosmos with positive curvature is given based on its projective structure. The completed compactified manifold is a projective space P^4. The cosmologic singularity is a S^3 bounding the non-orientable space-time, which is a four-dimensional Möbius bubble enclosed in a solid ball of four-dimensional time. Complex extension of the manifold leads to a Kähler metric on CP^4, which is invariant under $U4$.

INTRODUCTION

The first modern scientist who suggested a singular genesis of the cosmos was George Lemaître [1931a, 1933, 1950, 1958; Dirac 1968; Peebles 1984; North 1965; Kerszberg 1989]. It would be fitting to name the singularity after him superceding the pejorative terminology introduced by Fred Hoyle. This space-time singularity raised at once the question of its prehistory. Here Lemaître tried to conjure the spectre of quantum cosmology by referring to the uncertainty principle. He wrote: "The whole story of the world need not have been written down in the first quantum like a song on a disc of a phonograph" [Lemaître 1931b].

When Richard Tolman studied the radiation cosmos [Tolman 1934] he suggested a periodic time-dependence as a solution to the pre-history problem. He also speculated about entropy in a cyclic universe [Davis 1974].

In the 1950's a singular origin of the universe gained credibility through the studies of George Gamow and his collaborators [Gamow 1952; Alpher and Herman in Reines 1972] when theories of stellar evolution together with observations of globular clusters and radioactive dating established a finite age for the cosmos.

Simultaneously, Einstein's theory of gravitation was scrutinized from the point of view of modern mathematics by André Lichnerowicz [1955] and many of his followers. The singularities were excised from the space-time manifold and continuations through a singularity - a la Tolman - deemed pointless.

In 1965 Arno Penzias and Robert Wilson announced the discovery of the microwave background [Penzias 1972] thus strengthening the evidence for a singular genesis. In the same year Roger Penrose published his singularity theorem [Penrose 1965], which made it clear that the Lemaître singularity was generic and not the artifact of the assumed high spatial symmetry of the Friedmann model.

During the last quarter century a singular beginning of the universe has become widely accepted although the sign of the spatial curvature for the cosmos has not been determined. On the theoretical side the nature of the Lemaître singularity (short "the Lemaître ") has not been established. Singularities have been analyzed especially by Roger Penrose, Robert Geroch, and Bernd Schmidt [Penrose in Lebovitz 1978, Penrose and Rindler 1984, Geroch and Horowitz in Hawking and Israel 1979, Schmidt 1971, 1972, 1974, Hawking and Ellis 1973, Naber 1988]. However, in the Lemaître case no unique extension of the space-time has emerged.

Attempts by us to use rigid isometric embedding of the Friedmann manifold into 5-dimensional Minkowski space [Lauro 1983, Lauro and Schucking 1985, Wang 1986] did not achieve the desired extension to a larger manifold.

Since the metric and its conformal structure become singular at the Lemaître one may ask whether there are deeper structures in the geometry that can carry one further. The importance of a local projective structure as defining the laws of inertial motion was investigated by Juürgen Ehlers, Felix Pirani, and Alfred Schild [O. Raifeartaigh 1972, Ehlers 1977] and it is in fact a projective structure that we shall use to extend the Tolman cosmos.

It was principally through the work of Felix Klein - helped by Sophus Lie - [Klein 1893; Birkhoff and Bennet in Aspray and Kitcher 1988; Yaglom 1988] that the role of the metric in geometry was clarified. Klein showed [Klein 1928; Busemann 1953], generalizing Arthur Cayley's ideas, that Euclidean and non-Euclidean geometries in n dimensions can be obtained as specializations of projective geometry. He defined distance d between two points x and y by means of a quadratic form

$$<x,y> = <y,x> \qquad (0.1)$$

in (n+1) variables as

$$d(x,y) = c \ln \frac{<x,y> + (<x,y>^2 - <x,x><y,y>)^{\frac{1}{2}}}{<x,y> - (<x,y>^2 - <x,x><y,y>)^{\frac{1}{2}}} \qquad (0.2)$$

where c is a constant and

$$<x,y> = a_{AB} x^A y^B, \qquad a_{AB} = a_{BA}, \qquad A, B = 0, 1, ..., n. \qquad (0.3)$$

The coordinates x^A, y^B are homogeneous coordinates.

It is somewhat simpler to formulate this idea from the differential point of view by considering Kähler metrics on the complex n-dimensional space CP^n [Kähler 1933; Chern 1979]. Here the Study-Fubini metrics [Fubini 1903; Study 1905; Goldberg 1962] are simply obtained from the Kähler potential φ

$$\phi = c \ln a_{AB} z^A \overline{z^B}, \qquad a_{AB} = \overline{a}_{BA}, \qquad \det|a_{AB}| \neq 0, \qquad (0.4)$$

with a Hermitian form a_{AB} and real c. For the real projective space the construction works analogously.

By using Kähler potentials, which are - up to gauge transformations - homogeneous of degree zero, we obtain now an interesting class of metrics on projective spaces that generalize in the real case the spaces of constant curvature and in the complex the Study-Fubini metrics. It turns out that the Tolman cosmos - with a slight adjustment to its topology - belongs to these metrics like other Friedmann models.

The topological modification arises as follows. When we introduce homogeneous coordinates into a time-symmetric model with an extremal radius at t = 0 it is suggestive to consider this S^3 as a projective hyperplane "at infinity". This forces us to identify opposite points on the S^3 to give it the correct topology for a projective space. By continuity one is then led to an elliptic topology for the Tolman cosmos. The choice of the other coordinates is dictated by the O_4 - symmetry of the metric. The use of t as one of the five homogeneous coordinates (instead of some function of t) derives from the assumed Kähler property of the metric. The global projective structure consists of charts which cover the manifold and are related by linear transformations in homogeneous coordinates.

The desire to extend space-time beyond the Lemaître is the deeply felt wish to view the cosmos as an organic whole. Mathematicians have found such pleasing wholeness in the analytical character of solutions to problems involving elliptic operators. Pascual Jordan [Jordan 1952] illustrated the nature of this wholeness: "It is already the same type of conclusion when we see a live cat's paw emerging from a hole and we infer from it the existence of an attached cat".

It is the holistic structure of projective Kähler space which we see in Tolman's cosmos that encourages us to extend it beyond the Lemaître .

1. THE TOLMAN COSMOS

The Tolman cosmos [Tolman 1934] has often been used to characterize the structure of space-time close to its initial singularity. This model is a solution of the Einstein-Hilbert equations

$$G^\mu_{\ \nu} \equiv R^\mu_{\ \nu} - \frac{1}{2} \delta^\mu_{\ \nu} R = -\kappa T^\mu_{\ \nu}, \quad \mu, \nu = 0, 1, 2, 3 \qquad (1.1)$$

for the metric

$$ds^2 = g_{\mu\nu} dx^\mu dx^\nu \equiv -dt^2 + R^2(t) d\sigma^2. \qquad (1.2)$$

Here R(t) - different from the Ricci scalar R in (1.1) - is the positive radius of a 3-space with spatially constant positive curvature. The 3-dimensional metric $d\sigma^2$ can be given by

$$d\sigma^2 = d\hat{x}^\alpha d\hat{x}^\alpha, \quad \hat{x}^\alpha \hat{x}^\alpha = 1, \quad \alpha = 1, 2, 3, 4, \qquad (1.3)$$

using four dependent coordinates $\hat{x}^\alpha$.

The left hand side of (1.1) becomes then

$$G^0_{\ 0} = \frac{3(1+\dot{R}^2)}{R^2}, \quad G^j_{\ k} = \frac{\delta^j_k (1+\dot{R}^2 + 2R\ddot{R})}{R^2}, \quad G^j_{\ 0} = G^0_{\ j} = 0 \qquad (1.4)$$

where indices j and k run from one to three and $\dot{R} = dR/dt$, $\ddot{R} = d^2R/dt^2$.

On the right hand side of (1.1) we have, with c = 1, the relativistic gravitational constant $\kappa = 8\pi$ times Newton's gravitational constant G. The energy-momentum tensor $T^\mu{}_\nu$ describes a spatially homogeneous and isotropic radiation field. This tensor is assumed to be symmetric and traceless with a vanishing divergence. Because of its homogeneity and isotropy - the action of the group $O_4(R)$ - we have then

$$-\kappa T^0{}_0 = \frac{3a^2}{R^4}, \quad -\kappa T^j{}_k = -\frac{a^2}{R^4}, \quad T^0{}_k = T^j{}_0 = 0, \quad a = const. > 0. \quad (1.5)$$

If we count time t from the instant when R(t) takes its maximum we have as solution of (1.1)

$$R^2 = a^2 - t^2, \quad (1.6)$$

and the constant a becomes this maximal radius. At $t = \pm a$ the energy density $T^0{}_0$ becomes infinite and so the eigenvalues of the Ricci-tensor too.

These singularities restrict the validity of the solution to the range $|t| < a$. In this range the Tolman model T is a simply connected orientable and time-orientable analytic manifold with topology $S^3 \times R$. T can be covered through a stereographic atlas for S^3 with two analytically related charts. The manifold T is, clearly, incomplete since time-like geodesics like $\hat{x}^\alpha = const.$ reach the singularities at finite proper time. The metric of T is invariant under the group $O_4 \times C_2$, where C_2 is the 2-element group of time reversals ($t \to \pm t$).

The metric is also conformally invariant under a map with the generator

$$X = R(t) \frac{\partial}{\partial t}. \quad (1.7)$$

To simplify the subsequent formulae we shall put a = 1. The general case can then be recovered by a scale transformation: $t \to t/a$, $R \to R/a$, $ds \to ds/a$, $T^\mu{}_\nu \to a^{-2} T^\mu{}_\nu$.

2. THE EXTENSION PROBLEM

We shall assume in the following that all manifolds have no boundaries and that they are connected, Hausdorff, and para-compact [Geroch 1979]. We add to this catalogue the assumption that space-time (and its extension) is an analytic manifold. This does - by no means - imply that space-time functions, like the components of fields and metrics, are analytic. But without it the notion of an analytic space-time variable would be meaningless.

The analyticity condition allows us also to restore a manifold if a mischievous soul should punch holes into a chart declaring them "out of bounds" (singularities). If the chart remains path-connected the analytical identity map will then repair the damage. If physical reasons for weakening this requirement should become known one might try to weaken the analyticity requirement accordingly.

Analyticity alone is not much help for an extension of the model T. One could think of T as a S^4 with North and South pole removed and extend it analytically into the S^4 and plugging these two points back in where the metric becomes singular. But one might equally well extend T analytically to the cylinder $S^3 \times R$ or $S^3 \times S$ without impairing the $O_4 \times C_2$ symmetry of the metric.

The metric, becoming singular for $t^2 = 1$, is of no avail for the extension of the manifold. The conformal transformations cannot push us beyond this boundary either because according to (1.7) their generator vanishes there. We have to look for deeper structures to effect a continuation of T.

As we learned from Felix Klein's Erlanger Programm [Klein 1893] the structure of space can be perceived as a hierarchy characterized by the action of finite dimensional Lie groups the largest of which is the projective one $PSL_{n+1}(R)$ acting on the PR^n. It is this group that we shall employ for an extension of the Tolman universe.

3. THE PROJECTIVE TOLMAN COSMOS

It is a remarkable fact that the Tolman cosmos can be seen as the G-space of the Lie group germ of $PSL_5(R)$. For this to become possible we have to introduce a slight, but far reaching, modification of the model T, which we shall call the model T': the projective Tolman cosmos. In the model T' we identify events with the coordinates

$$\{ t, \hat{x}^\alpha \} \longleftrightarrow \{ -t, -\hat{x}^\alpha \}, \tag{3.1}$$

The manifold T' will then become the "equatorial belt" of the projective space $P^4(R)$. Big bang and big crunch will thus be identified. Such a model for the de Sitter universe had been suggested by Schrödinger [1956].

We describe T' now by means of five charts (λ) and (o) where the chart index(λ) takes the values 1, 2, 3, and 4. The four times five coordinates of these charts are defined in terms of t and $\hat{x}^\lambda$ by

$$x^0_{(\lambda)} = \frac{t}{\hat{x}^\lambda}, \quad x^j_{(\lambda)} = \frac{\hat{x}^j}{\hat{x}^\lambda}, \quad j \neq \lambda \quad \text{and} \quad x^\lambda_{(o)} = \frac{\hat{x}^\lambda}{t}. \tag{3.2}$$

The charts (λ) and (μ) are related by

$$x^0_{(\lambda)} = \frac{x^0_{(\mu)}}{x^\lambda_{(\mu)}}, \quad x^j_{(\lambda)} = \frac{x^j_{(\mu)}}{x^\lambda_{(\mu)}}, \quad j \neq \lambda, \quad x^\mu_{(\lambda)} = \frac{1}{x^\lambda_{(\mu)}}. \tag{3.3}$$

The charts (λ) and (o) are related by

$$x^0_{(\lambda)} = \frac{1}{x^\lambda_{(o)}}, \quad x^j_{(\lambda)} = \frac{x^j_{(o)}}{x^\lambda_{(o)}}, \quad j \neq \lambda. \tag{3.4}$$

Here, and in the following, indices j, k, l take the three values in the set {1,2,3,4} different from λ. The metric takes in these charts the forms

$$ds^2 = -\frac{\left[dx^0_{(\lambda)} - \frac{x^0_{(\lambda)} x^j_{(\lambda)} dx^j_{(\lambda)}}{1 + x^k_{(\lambda)} x^k_{(\lambda)}} \right]^2}{(1 + x^l_{(\lambda)} x^l_{(\lambda)})^2} + \left[1 - \frac{(x^0_{(\lambda)})^2}{1 + x^k_{(\lambda)} x^k_{(\lambda)}} \right] \left[\frac{dx^j_{(\lambda)} dx^j_{(\lambda)}}{1 + x^k_{(\lambda)} x^k_{(\lambda)}} - \frac{(x^j_{(\lambda)} dx^j_{(\lambda)})^2}{(1 + x^k_{(\lambda)} x^k_{(\lambda)})^2} \right],$$

$$\tag{3.5}$$

$$ds^2 = \left[1 - \frac{1}{x^\lambda_{(o)} x^\lambda_{(o)}} \right] \frac{dx^\lambda_{(o)} dx^\lambda_{(o)}}{x^\mu_{(o)} x^\mu_{(o)}} - \frac{(x^\lambda_{(o)} dx^\lambda_{(o)})^2}{(x^\mu_{(o)} x^\mu_{(o)})^2}. \tag{3.6}$$

The last, the (o)-chart, is of particular interest. It exhibits clearly the $O_4(R)$ symmetry of the metric of T'. The coordinate singularity at $x^\lambda_{(o)} x^\lambda_{(o)} = \infty$ is the 3-dimensional elliptic space covered by the four charts with

$$x^0_{(\lambda)} = 0 \qquad (3.7)$$

which is a hyperplane at infinity in the $P^4(R)$ and corresponds in the $\{t, \hat{x}^\alpha\}$ coordinates to the space of maximal radius $t = 0$.

The rest of T' is the exterior of the S^3

$$x^\lambda_{(o)} x^\lambda_{(o)} > 1. \qquad (3.8)$$

All four coordinates $x^\lambda_{(o)}$ are space-like in this region. The old time coordinate t^2 is given by

$$t^2 = \frac{1}{x^\lambda_{(o)} x^\lambda_{(o)}}. \qquad (3.9)$$

The solid ball

$$x^\lambda_{(o)} x^\lambda_{(o)} \leq 1. \qquad (3.10)$$

complements T' to PR^4. The singularity of the metric occurs now on the S^3

$$x^\lambda_{(o)} x^\lambda_{(o)} = 1. \qquad (3.11)$$

The unique extension of the manifold T' can be carried out in a number of ways. We can use analyticity of the identity map on the (o)-chart to plug the hole or we can use projective transformations by extending the Lie group germ of $PSL_5(R)$ into its full group and letting this group act on T'.

There is still another somewhat circumstantial way not without interest: By complexification one can go from T' to the holomorphic manifold CT' which can be uniquely extended into the simply connected holomorphic manifold CP^4 whose real part is PR^4, the extension of the manifold T'.

It should be mentioned that T' is neither orientable nor time-orientable. The extended manifold PR^4 is also not orientable.

4. THE METRIC EXTENSION

The metric in the (o)-chart (3.6) extends the metric from T' into the region

$$x^\lambda_{(o)} x^\lambda_{(o)} > 0. \qquad (4.1)$$

We have excluded here the point $x^\lambda_{(o)}$ since the metric becomes singular there. At the cosmological singularity, that occurs at

$$x^\lambda_{(o)} x^\lambda_{(o)} = 1 \qquad (4.2)$$

the induced metric on the S^3 vanishes completely. It is even analytic here but the Ricci tensor becomes infinite since it makes use of the inverse metric. This poses the question of the uniqueness of the extension of the metric into the region

$$x^\lambda_{(o)} x^\lambda_{(o)} < 1. \qquad (4.3)$$

The field equations

$$G^\mu_{\ \nu} + \kappa T^\mu_{\ \nu} = 0 \tag{4.4}$$

cannot be used because they become singular.

However, we shall use a slightly different version of the equations given by

$$g^{\frac{2}{3}} [G^\mu_{\ \nu} + \kappa T^\mu_{\ \nu}] = 0 \tag{4.5}$$

where g is the determinant of the metric. This leads to source terms no longer dependent on R according to (1.5). By assuming O_4-symmetry also for the continuation we can use equations (1.4) and (1.5) and get the two equations

$$(1+\dot{R}^2) R^2 - a^2 = 0, \quad (1+\dot{R}^2 + 2R\ddot{R}) R^2 + a^2 = 0. \tag{4.6}$$

The second equation is a consequence of the first for $\dot{R} \neq 0$, which applies here. We introduce new variables by

$$R^2 \equiv y, \quad t^2 \equiv x, \quad y' \equiv \frac{dy}{dx}. \tag{4.7}$$

The first equation becomes then

$$y + xy'^2 - a^2 = 0, \tag{4.8}$$

which is solved by

$$y = a^2 - x \tag{4.9}$$

with the initial conditions

$$y(a^2) = 0, \quad \frac{dy(a^2)}{dx} = -1. \tag{4.10}$$

We have from (4.8) that

$$y' = -\sqrt{\frac{a^2 - y}{x}} \tag{4.11}$$

fulfills the Lipschitz conditions for $x > a^2$. Thus

$$R^2 = a^2 - t^2 \tag{4.12}$$

is the unique solution.

We have thus for the metric of the extended Tolman model T'

$$ds^2 = -dt^2 + (a^2 - t^2) d\hat{x}^\alpha d\hat{x}^\alpha, \quad \hat{x}^\alpha \hat{x}^\alpha = 1. \tag{4.13}$$

The cosmological singularity at $t^2 = a^2$ is caused by a vanishing of the metric on a S^3. For $t^2 > a^2$ the metric becomes totally time-like. We arrive thus at the following picture for the structure of the universe:

The space-time manifold T' is a bubble in a four-dimensional infinite ocean of sheer time. This engulfing ocean, the "chronosphere", encloses the continental bubble by a shore which is a hypersphere. This shore is the moment of creation and destruction of the bubble where three time-like dimension in the chronosphere turn into three space-like dimensions and back.

This picture is beautiful and simple.

5. ISOMETRIC EMBEDDING

Putting a^2 again equal to one we can write the metric (1.2) as

$$ds^2 = -dt^2 - t^2 d\sigma^2 + d\sigma^2, \qquad d\sigma^2 = d\hat{x}^\alpha \, d\hat{x}^\alpha, \qquad \hat{x}^\alpha \hat{x}^\alpha = 1. \tag{5.1}$$

This suggests the following embedding. We take a seven-dimensional space which is the direct and metrical product of a negative definite Euclidean R^4 and three dimensional projective space P^3 with its invariant positive definite metric. The R^4 will be described by four projective coordinates x^α which may not all vanish simultaneously. The metric of the seven-dimensional space is thus

$$ds^2 = -dt^\alpha dt^\alpha + \frac{dx^\beta dx^\beta}{x^\gamma x^\gamma} - \frac{(x^\beta dx^\beta)^2}{(x^\gamma x^\gamma)^2}, \qquad \alpha, \beta, \gamma = 1, 2, 3, 4. \tag{5.2}$$

We consider in this space the algebraic manifold with equations

$$t^\alpha x^\beta - t^\beta x^\alpha = 0. \tag{5.3}$$

These equations have the solutions

$$t^\alpha = 0, \qquad x^\beta \text{ arbitrary} \tag{5.4}$$

and

$$x^\alpha = \lambda t^\alpha, \qquad \lambda \neq 0, \qquad t^\alpha t^\alpha > 0. \tag{5.5}$$

The first solution gives the P^3 for $t = 0$ at maximal expansion. The others describe the rest of P^4 except the compactification point at $t^2 = \infty$. To see that the embedding is isometric one introduces polar coordinates in the "space" of the t^α writing

$$-dt^\alpha dt^\alpha = -dt^2 - t^2 d\sigma^2. \tag{5.6}$$

If we drop two dimensions the metric (5.2) becomes

$$ds^2 = -(dt^1)^2 - (dt^2)^2 + d\phi^2. \tag{5.7}$$

where the manifold is given by

$$\phi = \tan^{-1}(t^2/t^1), \qquad -\frac{\pi}{2} \leq \phi \leq \frac{\pi}{2}. \tag{5.8}$$

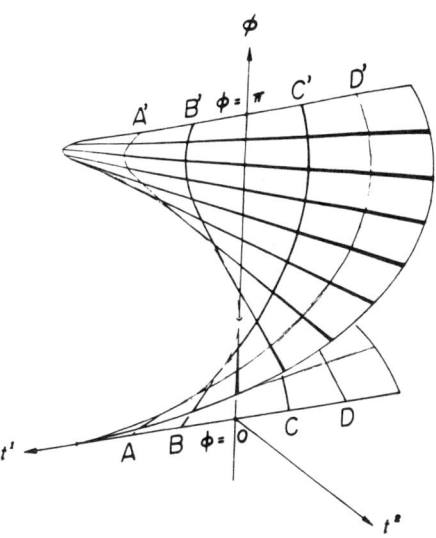

Fig. 1. The pseudo-right helicoid with equation $\tan \phi = t^2/t^1$ is isometrically embedded into a space with metric $ds^2 = (d\phi)^2 - (dt^1)^2 - (dt^2)^2$. For the two-dimensional model of T' the two generators given by the t^1 axis through the points $\phi = 0$ and $\phi = \pi$ on the ϕ-axis have to be identified such that ($A = A'$, $B = B'$, etc.).

The ϕ-axis represents space at its largest extension for $t = 0$. Time $|t|$ increases radially from the ϕ-axis. The helical arc BC' which is null represents the big bang, the other CB' the big crunch. Together they describe a circle. The Möbius band inside this circle represents the Tolman space-time T'. The region outside, e.g. at A and D shows part of the chronosphere. The generators $\phi = $ const. represent the idealized world lines of galaxies at rest with respect to the microwave background.

The upper and lower edges of this right helicoid have to be identified. This gives a Möbius strip. The strip about the axis (see Fig.1)

$$0 \leq (t^1)^2 + (t^2)^2 < 1 \tag{5.9}$$

represents the model T' with one space dimension. The cosmological singularity is the S^1 with

$$(t^1)^2 + (t^2)^2 = 1 \tag{5.10}$$

while the chronosphere with two-dimensional time is given by

$$(t^1)^2 + (t^2)^2 > 1. \tag{5.11}$$

Compactification by a point at infinite $|t|^2$ gives the manifold the topology of $P^2(R)$.

The generators of the helicoid (the horizontal time-like geodesics) pass serenely through the singularities. They would be the idealized world lines of galaxies at rest with respect to the microwave background. They came from the depth of the chronosphere and return to it after a sojourn through space-time.

The picture makes also clear that a global arrow of time does not exist.

6. THE COMPLEX EXTENSION

After extending T' to $P^4(R)$ one might as well consider its universal covering manifold S^4 as the extension of the original model T, undoing the event-identifications (3.1). Observation will have to determine which - if any of these or other - topologies apply to the actual cosmos.

Here, we want to point out, that while the difference between models T and T' is slight in the real it becomes profound upon complexification. The model T' extends into CP^4 and its metric in the (o)-chart becomes the real part of the Kähler metric

$$ds^2 = \left[1 - \frac{1}{Z^\alpha_{(o)} \bar{Z}^\alpha_{(o)}}\right] \frac{dZ^\alpha_{(o)} d\bar{Z}^\alpha_{(o)}}{Z^\beta_{(o)} \bar{Z}^\beta_{(o)}} - \frac{|\bar{Z}^\alpha_{(o)} dZ^\alpha_{(o)}|^2}{(Z^\beta_{(o)} \bar{Z}^\beta_{(o)})^2}, \quad \alpha, \beta = 1, 2, 3, 4.$$

(6.1)

The $O_4(R)$ invariance of T' extends now to the U_4 invariance of the complex extension which is simply connected. The 3-spheres $t^2 = $ const. become the real parts of CP^3 which are Kähler manifolds of maximal symmetry.

The general Kähler metric in chart C^4 invariant under U_4 is given by a Kähler potential

$$\phi = \phi(Z^\alpha \bar{Z}^\alpha)$$

(6.2)

where ϕ is a real function of one variable. The metric tensor is then

$$g_{\alpha\bar{\beta}} = \frac{\partial^2 \phi}{\partial Z^\alpha \partial \bar{Z}^\beta}$$

(6.3)

By restricting the Kähler metric to its real part and identifying it with that of T' the function ϕ can be easily determined also for more general Friedmann models. It was this procedure that led us to consider the projective variant of the model T.

In a complex extension of the model T one extends the $O_4(R)$ group to $O_4(C)$. The non-compactness of this group creates problems and leads to metrics of lesser symmetry (12 versus 16 dimensions compared with U_4). A possible physical interpretation of these complex extensions as phase spaces for which the closed Kähler two- form provides a symplectic structure will be discussed at another occasion.

7. CONCLUSION

We have sketched above a mathematical model for the creation (and the end) of space-time. The principal point in its favor is its simplicity. Newton's Rule 1 of Philosophy in his "System of the World", the third part of "Principia" was: "Natura enim simplex est" (Nature, namely, is simple) [1687]. But simplicity is not proof. At present nothing is known about the topology of the universe and a discussion of the model proposed will not be a confrontation of experiment or observation versus theory but an inquiry of consistency with cherished principles of physics. It will be necessary to study more general models and analyse their stability. One also has to be more specific about the source terms. But there are some general remarks that can be made already.

We have here the model of a kind of "phase-transition" for the metric as a result of the classical field equations of gravitation. The universe acquires a pre- and post- history and its approximate (conjectured) isotropy and homogeneity have now to be seen as imposed by the symmetry of the chronosphere. The horizon syndrome for which inflation was touted as a cure would no longer appear life threatening to a universe. A time-like congruence threads smoothly through the eye of the singularity.

Since the early days of quantum mechanics it has been clear that this theory applies to microscopic systems and macroscopic ones at low temperature. Most physicists would agree that it is irrelevant for celestial mechanics. Its use for the largest possible system at the highest temperatures, i.e. the earliest universe, in a discussion of the creation appears unnecessary when one can deal with it along more classical lines.

We are still ill prepared to discuss the physics in the chronosphere. But many of its laws are insensitive to a change of signature and it will be a challenge to see a richer world in the fullness of time.

The authors gratefully acknowledge discussions with Jerome Epstein, O. Greengard, Marcel Reginatto, and Guihua Zhang.

REFERENCES

Aspray W. and Kitcher P., 1988, "History and Philosophy of Modern Mathematics," Univ. of Minnesota Press.
Busemann, H. and Kelly, P.J., 1953, "Projective Geometry and Projective Metrics," Academic Press, New York.
Chern, S.S., 1979, "Complex Manifolds," Springer, New York.
Davis, P.C.W., 1974, "The Physics of Time Asymmetry," University of California Press, Berkeley and Los Angeles. Also as paperback (1977).
Dirac, P.A.M., 1968, The Scientific Work of George Lemaître , *Pontificae Academiae Scientiarium Commentarii*, Vol II, 11:1-20.
Ehlers, J. and Kohler, E., 1977, *J. Math. Physics* 18:2014.
Fubini, G., 1903, *Atti Inst.* Veneto 6:501.
Gamow, G., 1961, "The Creation of the Universe," Viking, New York, pVII.
Geroch, R., 1979, "General Relativity," S.W. Hawking and W. Israel, ed., Cambridge University Press, Cambridge.
Goldberg, S., 1962, "Curvature and Homology," Academic Press, New York. Also reprinted by Dover.
Hawking, S.W. and Ellis, G.F.R., 1973, "The Large Scale Structure of Space-time," Cambridge University Press, Cambridge.
Hawking, S.W. and Israel, W., 1979, "General Relativity," Cambridge Univ. Press, Cambridge.
Jordan, P., 1952, "Schwerkraft und Weltall," Vieweg. Braunschweig, p88.
Kähler, E., 1933, *Abh. Math. Sem.*, Hamburg, 9:173.
Kerszberg, P., 1989, "The Invented Universe," Clarendon Press, Oxford.
Klein, F., 1893, *Math. Annalen* 43:63. Reprint of 1872 Erlanger Programm.
Klein, F., 1928, "Vorlesungen über Nicht-Euklidische Geometrie," Springer, Berlin.
Lauro, S., 1983, PhD thesis, NYU.
Lauro, S. and Schucking, E.L., 1985, *Int. J. Theor. Phys.*, 24:367.
Lebovitz, N. et al. eds., 1978, "Theoretical Principles in Astrophysics and Relativity," University of Chicago Press, Chicago, Illinois.
Lemaître , G., 1931a, The Expanding Universe, *Monthly Not. Roy. Astr. Soc.,* 91:490-501.
Lemaître , G., 1931b, The Beginning of the World from the Point of View of Quantum Theory, *Nature,* 127:706.
Lemaître , G., 1933, L'univers en Expansion, *Ann. Soc. Sci. Brux.*, 53(A):51 - 85.
Lemaître , G., 1950, "The Primeval Atom," Van Nostrand, New York.
Lemaître , G., 1958, Rencontres avec Einstein, *Rev. Quest. Sci.,* 129-32.
Lichnerowicz, A., 1955, "Theories Relativistes de la Gravitation et de l'Electromagnetisme," Masson et Cie, Paris.

Naber, G., 1988, "Spacetime and its Singularities, an Introduction," Cambridge Univ. Press, Cambridge.
North, J., 1965, "The Measure of the Universe," Clarendon Press, Oxford.
O'Raifeartaigh, 1972, "General Relativity," Clarendon Press, Oxford.
Peebles, P.J., 1984, Impact of Lemaître 's Idea on Modern Cosmology, *in*: "The Big Bang and George Lemaître," A. Berger, ed., Reidel, Dordrecht, 23-30.
Penrose, R., 1965, *Phys. Rev. Lett.,* 14:57.
Penrose, R. and Rindler, W., 1984, "Spinors and Space-time," Vol. 1 and 2, Cambridge Univ. Press, Cambridge.
Penzias, A., 1972, "Cosmology, Fusion, and Other Matters," George Gamow Memorial Volume, F. Reines, ed., Colorado Associated Univ. Press. See also Chapter One by R.A. Alpher and R. Herman and literature there.
Schmidt, B.J., 1971, *J. Gen. Rel. and Grav.*, 1:169.
Schmidt, B.J., 1972, *Comm. Math. Phys.*, 29:49.
Schmidt, B.J., 1974, *Comm. Math. Phys.*, 36:73.
Study, E., 1905, *Math. Annalen*, 60:321.
Tolman, R.C., 1934, "Relativity, Thermodynamics and Cosmology," Oxford Univ. Press, Oxford, England, Reprinted by Dover Publications, Inc., New York 1987.
Wang, J.-Z., 1986, PhD thesis, NYU.
Yaglom, I.M., 1988, "Felix Klein and Sophus Lie," Birkhäuser, Boston.

SOME IDEAS ON THE COSMOLOGICAL CONSTANT PROBLEM

G. Veneziano

Theory Division, CERN
1211 Geneva 23, Switzerland

1. Outline

In this talk I shall report on some ideas for explaining the experimental smallness of the cosmological constant (CC).

I will start by recalling what the CC problem is and why its resolution cannot be a simple one. I shall then proceed with a review of the Baum-Hawking-Coleman (BHC) wormhole-based solution, stressing general features as opposed to technical details. Next I will criticize the BHC approach, by arguing that the vanishing of the CC is closely related, in that scheme, to a wormhole-induced quantum instability of the theory.

In the second part of the talk, I shall present two alternative ways of suppressing, rather than cancelling completely, the CC. The first approach is still based on topological effects, such as wormholes, while the second one is more conventionally based on perturbative quantum gravity corrections. Unfortunately, both schemes are still affected by (serious?) difficulties. They do contain, however, some new ideas which, I believe, are well worth pursuing.

2. The cosmological constant problem

The CC problem goes back to the very early days of General Relativity. In 1917, Einstein, searching for static solutions of his equations (the expansion of the Universe was not known at the time), introduced [1,2] an extra term committing what he later described as "the biggest blunder of my life".

Denoting by G and Λ the Newton and the cosmological constant, respectively, Einstein's modified equations take the form:

$$-R_{\mu\nu} + 1/2 g_{\mu\nu} R = \Lambda g_{\mu\nu} + 8\pi G T_{\mu\nu} \qquad (2.1)$$

The above equations also follow from the action:

$$\Gamma_{eff} = (16\pi G)^{-1} \int dx \sqrt{g}(2\Lambda - R) +$$

$$+\text{matter} + \text{higher derivative terms} \qquad (2.2)$$

We have denoted the action by Γ_{eff} in order to stress that we do not assume Einstein's theory to be fundamental. All we are assuming is that the effective, large distance theory of gravity is well described by the action (2.2). Indeed, this is the most general form of *local* action consistent with the invariance principles of General Relativity.

Equation (2.1) admits homogeneous, isotropic solutions, known as Friedmann-Robertson-Walker universes whose line element (metric) reads:

$$ds^2 = -dt^2 + a^2(t)\left[\frac{dr^2}{1 - kr^2} + r^2 d\Omega\right] \qquad (2.3)$$

where $k = 0, 1, -1$ correspond, respectively, to a flat, closed, open Universe.

Inserted into (2.1) this ansatz yields the standard cosmological equations for the scale factor $a(t)$:

$$(\ddot{a}/a) = \frac{\Lambda}{3} - \frac{4\pi G}{3}(\epsilon + 3p)$$

$$H^2 = (\dot{a}/a)^2 = -\frac{k}{a^2} + \frac{1}{3}\Lambda + \frac{8\pi G}{3}\epsilon \qquad (2.4)$$

where H is the Hubble "constant" and ϵ and p are the energy and pressure density respectively (defined by $T_0^0 = -\epsilon, T_1^1 = T_2^2 = T_3^3 = p$). Obviously, the cosmological term corresponds to a "vacuum energy density" $\epsilon_v = \frac{\Lambda}{8\pi G}$ with an equal and opposite pressure. From Eqs. (2.4) one can easily see:

i) the absence of a static solution if $\Lambda = 0$ and $\epsilon, p > 0$. This is the original Einstein observation[1].

ii) the fact that, if $\Lambda > 0$ and matter contributions can be neglected, $a(t)$ grows exponentially at large t:

$$\begin{array}{ll} a(t) = a(0)exp(Ht); & H^2 = \frac{\Lambda}{3}, k = 0 \\ a(t) = a(0)cosh(Ht); & H^2 = \frac{\Lambda}{3}, k = 1 \end{array} \qquad (2.5)$$

These represent so-called inflationary epochs of the Universe.

There is a variety of good reasons [3] for an inflationary period occurring in the early Universe. This makes the CC problem even more acute since we would like to have a large, positive CC in the early Universe which magically relaxes at later times to the present tiny value. This is constrained by [2]:

$$\Lambda(now) \lesssim H^2(now) \simeq 10^{-85} GeV^2, i.e.,$$
$$\epsilon_v \lesssim 10^{-47} GeV^4, \quad \text{or, finally} \qquad (2.6)$$
$$G\Lambda \lesssim 10^{-122}$$

To get such a tiny value of ϵ_v today, things should conspire extremely well, since any phase transition (e.g., the spontaneous breaking of gauge or global symmetries) is always accompanied by a change in the vacuum energy which is [2] at the very least 40 orders of magnitude larger than the bounds (2.6).

Needless to say, it is very hard to explain this magic cancellation from the "outside", i.e., from non-gravitational physics. Apart from antropic-principle explanations, the best chance seems to lie [2] in some mechanism provided by gravity itself and, more particularly, by quantum gravity effects.

3. The Baum-Hawking-Coleman solution

There are basically two ingredients to the BHC solution [4,5,6] of the CC problem. The first is technical (and even controversial): it consists of a set of recipes defining Euclidean Quantum Gravity [7]. It will be momentarily assumed. The second ingredient is more physical: it has to do with the possible appearance of non-localities in Γ_{eff} as a result of non-perturbative topological effects, e.g., of wormholes.

Let us give a qualitative description of wormholes [8]. These are metrics (in some theories they are even classical solutions) which have the characteristics of connecting together two smooth (e.g., flat) manifolds. Examples are depicted in Fig. 1 where, respectively, a wormhole joins together two flat Universes or connects two distant regions of the same Universe.

Let us consider the latter case in more detail. The effect of wormholes of characteristic size (width) ρ_W, is to induce, at scales much larger than ρ_W, a non-local correction to the usual local action (2.2) in the form:

$$\delta\Gamma_W = -1/2 \int dx \sqrt{g(x)} \int dy \sqrt{g(y)} \; C(x,y) \sim -1/2 CV^2 + O(V)$$

$$V \equiv \int dx \sqrt{g(x)} \qquad (3.1)$$

where one uses the important property that, for $|x - y| \gg \rho_W$, the wormhole contribution approaches a constant C. The order of magnitude of $C(\rho_W)$ is controlled by the wormhole

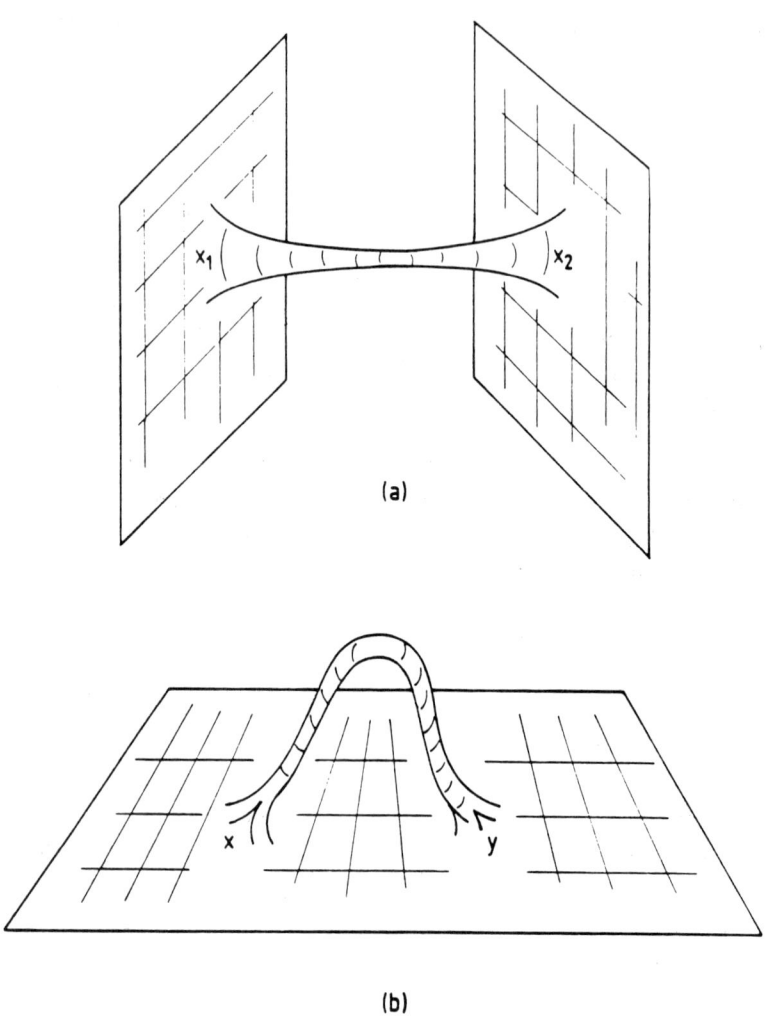

Fig. 1. A wormhole connecting (a) two flat Universes, (b) two distant parts of one Universe

action in the typical form:

$$C \simeq exp(-\frac{S_W}{\hbar}) \simeq exp(-3\pi(\frac{\rho_W}{\ell_P})^2), \quad \ell_P^2 = G\hbar \qquad (3.2)$$

For wormhole sizes an order of magnitude larger than the Planck length ℓ_P this is a very small number. We shall see, however, the way its effects can be as important as to kill, or drastically reduce, the CC.

The "proof", following Ref. [9], goes as follows. Consider the so-called partition function Z_{NL} associated with the non-local action:

$$\Gamma_{NL} = \Gamma_L + \delta\Gamma_W; \quad \Gamma_L(\Lambda) = (16\pi G)^{-1} \int dx \sqrt{g}(2\Lambda - R) \qquad (3.3)$$

where $\delta\Gamma_W$ has the form (3.1). Z_{NL} is defined by the (Euclidean) functional integral:

$$Z_{NL} = \int D[g_{\mu\nu}] exp(-\Gamma_{NL}) \tag{3.4}$$

One can get rid of the non-locality of the action by the simple trick of rewriting:

$$exp(-\delta\Gamma_W) = \int_{-\infty}^{+\infty} d\alpha \; exp\left(-\frac{\alpha^2}{2C} - \alpha V\right) \tag{3.5}$$

where we have momentarily assumed $C > 0$.

After exchange of the order of integration, Z_{NL} becomes a superposition of local-action partition functions with different values for Λ

$$Z_{NL} = \int d\alpha \; exp\left(-\frac{\alpha^2}{2C}\right) Z_L(\Lambda + 8\pi G\alpha);$$

$$Z_L(\Lambda) \equiv \int D[g] \; exp(-\Gamma_L(\Lambda)) \tag{3.6}$$

The question is: which values of $\Lambda_{eff} = \Lambda + 8\pi G\alpha$ dominate the integral? One can get easily an answer by assuming that the integral over the metrics in (3.6) is dominated by the saddle point corresponding to the classical solution:

$$R_{\mu\nu} = \Lambda_{eff} g_{\mu\nu} \tag{3.7}$$

On this classical solution the local action takes the value

$$\Gamma_L = -\Lambda_{eff}(8\pi G)^{-1} \int dx \sqrt{g(x)} \tag{3.8}$$

It is clear that negative values of Λ_{eff} are disfavoured. For positive Λ_{eff} the minimal action solution is known to be the 4-sphere of radius $r = (\frac{3}{\Lambda})^{1/2}$ which gives:

$$\Gamma_L^{min} = -(8\pi G)^{-1}\Lambda_{eff}\frac{8}{3}\pi^2 \cdot r^4 = -\frac{3\pi}{G\Lambda_{eff}} \tag{3.9}$$

Thus finally:

$$Z_{NL} \simeq \int_{-\infty}^{+\infty} d\alpha exp\left(\frac{-\alpha^2}{2C} + \frac{3\pi}{G(\Lambda + 8\pi G\alpha)}\right) \tag{3.10}$$

is completely dominated by the value of α corresponding to $\Lambda_{eff} = 0^+$. One can similarly prove [9] that any expectation value computed with the NL action by the usual path integral coincides with the one computed with a local action and $\Lambda = 0$. This looks almost too good to be true!

4. Criticism

There have been several criticisms of the above procedure and conclusions, e.g., on the way of computing expectation values of observables in Quantum Gravity [10]. Here I shall concentrate on a different class of objections [11,12,13] related to the physical interpretation of the singularity of the integrand in Eq. (3.10).

The reason why one can be suspicious about the naive BHC procedure is that one is dealing with ill-defined functional integrals. Besides the well known difficulties with integrating over the conformal factor [7], there is an additional problem brought in by the wormholes. If the wormhole constant C is positive, as assumed in Eq.(3.5), then the non-local action is unbounded from below for very large spherical Universes. The usually invoked Wick rotation of the conformal factor [7] can be easily argued [13] to be ineffective for this new divergence of the functional integral.

It is precisely the infinity of the non-local action which induces the singularity of the integrand in Eq. (3.10). Trusting the physics of that peak is making sense of the runaway vacuum of a theory with unbounded action, typically a dangerous game.

In order to clarify this point let us consider a simpler example of a theory with unbounded action at large distances: Einstein's action with a negative CC. Using formal manipulations similar to the ones employed with wormholes, we write:

$$Z(\Lambda) = \int_\Lambda^\infty d\alpha \int D[g_{\mu\nu}] \left(\frac{V}{8\pi G}\right) exp\left((16\pi G)^{-1} \int dx \sqrt{g}(R - 2\alpha)\right)$$

$$\approx \int_\Lambda^\infty d\alpha exp\left(\frac{3}{G\alpha}\right) \langle\frac{V}{8\pi G}\rangle_\alpha \qquad (4.1)$$

where we have again used a saddle point to integrate over the metrics. We see that, if $\Lambda > 0, Z$ is dominated by $\alpha = \Lambda = \Lambda_{eff}$, while , if $\Lambda < 0$, the integral is dominated by $\alpha = \Lambda_{eff} = 0^+$. We have thus apparently shown that a negative Λ implies a vanishing Λ_{eff}! The conclusion is again too nice to be true and (obviously?) false. The error can be attributed to using a saddle point for integrating over the metrics and *not* for integrating over α. But this is precisely what we have done in the wormhole case!

In order to get further insight into the problem let us go back to Eq.(3.10), but looking now for saddle points in α. The saddle-point condition is simply[11,13]:

$$\alpha = -\frac{24\pi^2 C}{(\Lambda + 8\pi G\alpha)^2} \qquad (4.2)$$

which is a cubic equation in α and thus has either one or three real solutions. The discussion is simplified by the assumption that the wormwhole "coupling" C is small because of (3.2).

We shall take C small to mean:

$$C \ll \Lambda^3 (192\pi^3 G)^{-1} \tag{4.3}$$

In this case there are two kinds of solutions:

A) $\alpha \simeq -\dfrac{24\pi^2 C}{\Lambda^2} \ll \dfrac{\Lambda}{8\pi G} \implies \Lambda \simeq \Lambda_{eff}$

B) $\Lambda_{eff} = \Lambda + 8\pi G\alpha = \pm 8\sqrt{3}\pi^{3/2} \left(\dfrac{GC}{\Lambda}\right)^{1/2} \ll \Lambda$ \hfill (4.4)

The B-type saddles are only real for $\Lambda > 0$ (recall that, for the moment, $C > 0$). Obviously, these are the saddles which can be interesting for quenching Λ. What do these saddles mean physically? The answer is quite obvious if we go back to the non-local action we started from. Varying directly Γ_{NL} we obtain a non-local modification of Einstein's equations reading:

$$R_{\mu\nu} - 1/2 g_{\mu\nu} R = (-\Lambda + 8\pi GCV) g_{\mu\nu} \tag{4.5}$$

which has a positive CC solution:

$$R_{\mu\nu} = \Lambda_{eff} g_{\mu\nu} \tag{4.6}$$

with Λ_{eff} given by:

$$\frac{(\Lambda_{eff} - \Lambda)}{8\pi G} = -\frac{24\pi^2 C}{\Lambda_{eff}^2} \tag{4.7}$$

after expressing the volume itself through Λ_{eff}. We note that the equation for Λ_{eff} is again cubic: indeed, it coincides with (4.2) after we identify Λ_{eff} with $\Lambda + 8\pi G\alpha$.

The nice thing about having gone back to Γ_{NL} is that the saddle points now have a clear, classical interpretation. By plotting Γ_{NL} as a function of V we can also see (Fig. 2a) that saddle B corresponds to a local *maximum* of Γ_{NL}. We can also conclude that, at least for the saddle points, the Euclidean Quantum Gravity machinery was not really necessary.

What about the $\Lambda_{eff} = 0$ result of BHC? That result *cannot* be obtained by a classical argument. Instead, as is quite obvious from Fig. 2a, it corresponds to the fact that Γ_{NL} is equal to $-\infty$ at $V = \infty$ if $C > 0$! The BHC result thus comes from the assumption that, quantum mechanically, the classical (de Sitter or anti de Sitter) vacuum becomes unstable and rolls away to $V = \infty$ or $\Lambda_{eff} = 0$.

We also understand now the case of $\Lambda < 0$ and no wormholes presented earlier. There too we could envisage the possibility that quantum corrections drive us away from the classical vacuum into the configuration of infinite 4-volume and vanishing CC.

207

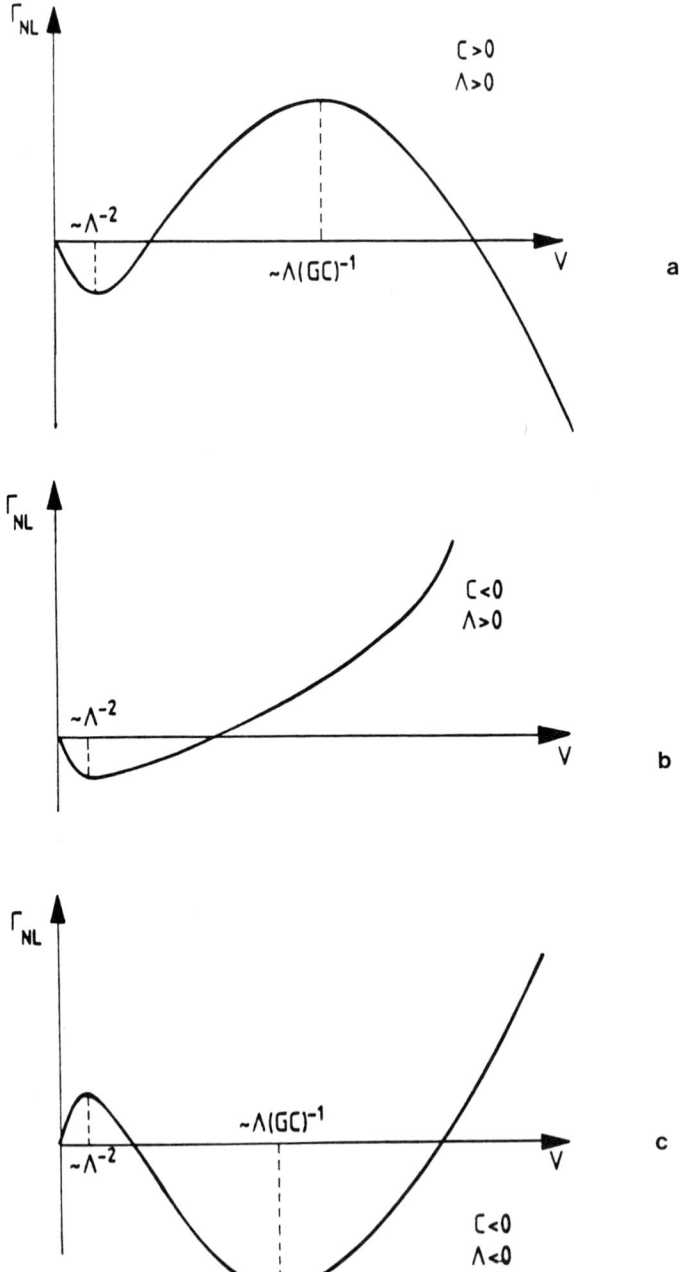

Fig. 2. Non-local Euclidean action as a function of the 4-volume V for three choices of the signs of C and Λ

Two attitudes are possible:

i) either one accepts theories with unbounded actions and exploits the resulting instabilities; or

ii) one works with cases in which the above instabilities are either absent or are avoided by some suitable definition of the quantum theory (i.e. of the path integral). The latter is the attitude generally adopted, for instance, in dealing with the conformal factor instability [7].

In the rest of this lecture I shall stick to this second, more conservative attitude without claiming, however, that the former is necessarily wrong. I only wish to add that other results obtained from Coleman's approach on natural constants other than Λ, crucially depend on taking seriously the infinite action instability and have no analog in the scenario that we shall now follow.

5. Quenching Λ by wormholes of the "wrong" coupling

Let us consider again the non-local action (3.3) but this time with a negative value for C. The situation can be analyzed as before either by introducing the auxiliary wormhole variable α, or, more directly, by looking for the stationary points of the non-local action. If we proceed in this second fashion we see immediately (Figs. 2b,c) that:

i) the non-local action is now bounded from below (at least in the direction of large, smooth manifolds) and

ii) the nature of the saddle points depends upon the sign of Λ. If $\Lambda > 0$ (Fig. 2b), there is only a real saddle corresponding to the usual de Sitter Universe with $\Lambda_{eff} = \Lambda$. Thus, in this case, the wormhole effect is negligible and inflation in the early Universe is maintained. If instead $\Lambda < 0$ (Fig. 2c), a new saddle appears (with lower Euclidean action) corresponding to:

$$\Lambda_{eff} = 8\sqrt{3}\pi^{3/2}(GC/\Lambda)^{1/2} \qquad (5.1)$$

This is precisely the saddle point described before, corresponding to a huge (for very small $|C|$) quenching of the CC. We shall discuss in a moment the conditions under which this quenching can be phenomenologically sufficient.

What happens in the wormhole-variable language for $C < 0$? In particular, how is the BHC effect avoided? Referring to [11] for details, I shall just mention that, for $C < 0$, the α integration should be done on the imaginary, rather than on the real axis (i.e., via a Fourier transform). The BHC singularity is now off the integration contour and does not cause any infinity for the partition function Z_{NL}. The latter is instead dominated by a saddle point which leads exactly to the predictions described above.

The situation is nicely summarized in Hawking's own words [13] as:

If $C > 0$, the integral over the metrics diverges; if $C < 0$, the integral over the α's diverges.

In the former case, the BHC instability takes place and, if one can make sense of it, the cosmological constant will vanish (and other constants of nature will be similarly determined too).

In the second case, the integral over the α's is simply to be rotated and everything is well defined. The BHC instability is avoided and so is, unfortunately, the prediction that $\Lambda_{eff} = 0$. It is still possible, however, to depress in this case a *negative* CC, through the saddle point (4.4).

In order to see under which conditions such a suppression is sufficient, let us rewrite (4.4) in the form:

$$G\Lambda_{eff} = (G\Lambda)^{1/2}(CG^2/\Lambda^2)^{1/2}(8\pi\sqrt{3}\pi^{3/2}) \tag{5.2}$$

From the experimental bound (2.6) we get immediately:

$$(CG^2/\Lambda^2) < 10^{-248}/G\Lambda \tag{5.3}$$

where we can expect $G\Lambda$ to lie anywhere in the range $10^{-80} - 1$. As explained at the beginning, the smallness of C is controlled by the minimal wormhole action, so that we may assume, as an order of magnitude:

$$(CG^2/\Lambda^2) = exp(-\hbar^{-1}S_W^{min}) = exp(-3\pi(\rho_W^{min}/\ell_P)^2) \tag{5.4}$$

It is easy to see that the experimental bound is satisfied for a minimal wormhole size 6 to 8 times larger than ℓ_P. We thus see that the success of the scheme depends on the short distance behaviour of quantum gravity or, if we prefer, on the nature of its short distance cutoff.

Probably quantum gravity does not make sense without an ultra-violet cut-off anyway. An example of a cut-off gravity theory is provided by string theory, which possesses a fundamental scale of its own:

$$\lambda_s = (\alpha\hbar)^{1/2} \tag{5.5}$$

If, as suggested from several points of view [14,15,16] , λ_s plays the role of a minimal observable scale in string theory, then it will also provide a minimal scale for topological fixtures such as wormholes. In this case, after use of the string-unification formula [17]

$$\lambda_s^2/\ell_P^2 = 16/\alpha_{GUT} \tag{5.6}$$

one obtains the amusing result:

$$exp(-\hbar^{-1}S_W^{min}) \simeq exp(-48\pi/\alpha_{GUT}) \tag{5.7}$$

which shows on one hand that the suppression of Λ_{eff} is exponential in the inverse fine structure constant (cf., instanton effects) and, on the other hand, that it is numerically sufficient for an α_{GUT} lying approximately in the experimentally favoured range.

Before claiming victory, however, we still have quite a few problems to address:

1. The sign of C that we need does not seem to be the one that follows from either intuitive arguments [18] or from model calculations[13].

2. Adding more wormhole-connected universes "dilutes" the quenching effect by the average number of mutually connected Universes [11]. In the simplest scenario this number is so huge that the final suppression is only logarithmic in C and this renders the hope of suppressing sufficiently Λ almost vain.

3. Finally, the whole scheme is based on the poorly understood physics of non perturbative effects in euclidean quantum gravity.

In view of the above, I shall turn, in the last part of this talk, to an alternative mechanism which, while employing similar ideas to the ones used until now, is based on the more conventional and better understood physics of perturbative radiative corrections as a way to induce a change in the classical vacuum of the theory, the well-known Coleman-Weinberg (the "other" Weinberg and, I should add, the "other Coleman's") mechanism.

6. Quenching Λ a-la-Coleman-Weinberg

Let us briefly recall the standard Coleman-Weinberg (CW) mechanism [19] for dynamical symmetry breaking. There one considers, for instance, a massless $\lambda(\phi^*\phi)^2$ theory coupled to electromagnetism. At tree level the scalar potential has a stable, absolute minimum at $\phi = 0$ and the gauge symmetry is unbroken. Under certain restrictions on the coupling λ and of the gauge coupling, CW argued that the one-loop corrected effective potential reliably describes quantum corrections to the tree-level classical picture and that the new scalar potential acquires a term:

$$V_1 = \beta(\phi^*\phi)^2 log(\phi^*\phi/\mu^2) \tag{6.1}$$

where $\beta > 0$ is a known function of the renormalized couplings and μ is the subtraction point. Obviously, V_1 becomes negative near $\phi = 0$ making the tree level vacuum unstable.

ϕ gets a non-zero expectation value and the gauge symmetry is broken through a dynamical Higgs mechanism (the so-called CW mechanism).

How can something like this work in the case of gravity and of the CC? Let us take the tree level action to be the usual:

$$\Gamma_{tree} = (16\pi G)^{-1} \int dx \sqrt{g(x)} (2\Lambda - R) \tag{6.2}$$

and let us try to guess a one-loop correction that might do the job. Later we shall see whether or not such a correction does come out of a bona fide calculation.

The toy-model one-loop correction to (6.2) is taken [20] to be of the form:

$$\Gamma_1 = -\gamma \Lambda^2 \int \sqrt{g(x)} log(\frac{\lambda_{UV}^2}{\lambda_{IR}^2}) dx \quad \gamma > 0 \tag{6.3}$$

where λ_{UV} is the UV cut-off needed to make sense out of quantum gravity calculations (e.g., λ_s in the case of string theory) while the much more important quantity λ_{IR} is a physical, $g_{\mu\nu}$ dependent IR cut-off that the theory itself should determine in terms of large distance physics. Our ansatz (6.3) is quite similar to the CW expression with λ_{IR} and λ_{UV} replacing $\phi^*\phi$ and μ^2 respectively. The fact that the one-loop correction to (6.2) has a logarithmic UV divergence proportional to the cosmological term with a coefficient proportional to Λ^2 is already known [21]. Thus our ansatz is not so wild... until we make an assumption on what one should write for λ_{IR}. The toy model of Ref. [20] is defined by the identification:

$$\lambda_{IR} = V^{1/4} \tag{6.4}$$

and thus has a total (non-local) action:

$$\Gamma_{eff} = \Gamma_{tree} - \gamma \Lambda^2 V \ log(\lambda_{UV}^2 V^{-1/2}), \quad \gamma > 0 \tag{6.5}$$

The stationary points of (6.5) are given by:

$$R_{\mu\nu} = \Lambda_{eff} g_{\mu\nu} \tag{6.6}$$

where:

$$\frac{\Lambda_{eff} - \Lambda}{8\pi G} = -\gamma G \Lambda^2 log(\lambda_{UV}^2 \Lambda_{eff}) \tag{6.7}$$

(6.7) is a transcendental equation for Λ_{eff} whose qualitative solutions are easily found graphically: in analogy with the wormhole case with $C < 0$, one finds that, for $\Lambda > 0$, $\Lambda_{eff} \simeq \Lambda$ while, for $\Lambda < 0$, Λ_{eff} is positive and suppressed, this time exponentially:

$$\Lambda_{eff} = \lambda_{UV}^{-2} exp(-1/G|\Lambda|) \tag{6.8}$$

This damping is quite sufficient for $\lambda_{UV} \geq \ell_P$ and $G|\Lambda| < 0(10^{-3})$, i.e., for an easy-to-achieve set of parameters.

How about the true one-loop calculation? This is not so trivial but can be done [22] (under standard assumptions). The one-loop effective action takes, as usual, the form of the *trlog* of the operator Ω of quadratic fluctuations around the chosen metric at which the effective action is computed. There are complications due to the necessity of introducing:

i) gauge fixing terms,

ii) the corresponding ghost fields and, finally,

iii) the De Witt-Vilkovisky [23] improvement needed to preserve the general co-ordinate invariance of the result.

The *trlog* Ω is then computed via the standard heat-kernel method as:

$$trlog\Omega = Tr \int_{\tau_{min}}^{\infty} d\tau \ exp(-\tau\Omega) = \int dx \sqrt{g(x)} \int_{\tau_{min}}^{\infty} d\tau \langle x|exp(-\tau\Omega)|x\rangle \qquad (6.9)$$

In this formulation the UV-cut-off dependence comes from the small τ region of integration after identification of τ_{min} with λ_{UV}^2. The small τ expansion of $< x|exp(-\tau\Omega)|x >$ is well known [21] and leads to a bunch of local operators multiplying different powers (or logarithms) of λ_{UV}.

As already argued, the IR cut-off should be intrinsic, i.e., related to the manifold on which we are performing the calculation. Obviously, the large τ behaviour is related to the spectrum of small eigenvalues of Ω, a rather "classical" mathematical problem. Somehow this spectrum depends on global properties of the manifold such as its size and shape. In the words of Mark Kāc [24] one is "hearing the shape of a drum!" This is how non-locality creeps into the evaluation of Γ_1. Note that the non-locality of quantum correction was already pointed out by Bryce De Witt as early as in 1967 [25].

The exact result is not known for general manifolds. It is, however, calculable for spheres and arguments can be given to extend the estimate to arbitrary, large and smooth enough manifolds for which boundary effects should not play a major role. The bottom-line result obtained in Ref.[22] reads:

$$\Gamma_{1 \ loop} \simeq \frac{3\Lambda^2}{8\pi^2} V \ log(\lambda_{UV}^2 M^2);$$

$$M^2 \equiv max\{|\Lambda|, V^{-1/2}\} \qquad (6.10)$$

which is *almost* what we had in the toy model (6.5). However, the difference between what we need and what we got is crucial. Even leaving out the fact that we got the "wrong" sign in front of the logarithm – something that could change, for instance, in supergravity – the disturbing fact remains that the argument of the *log* is not just V but contains the bare CC too. It is easy to see that, as a result, the logarithm can never grow large enough to appreciably change the tree level classical solution.

At a closer look, we see that the bare CC plays the role of IR cut-off much like a mass in ordinary theories. We thus lose the CW effect very much in the same way as CW would if they introduced a mass for the scalar field. This fact tells us, on one hand, that our perturbative approach may be doomed and, on the other, it may suggest a new possibility.

A positive (negative) CC looks indeed like a positive (negative) squared mass, not only in the way it regulates the IR but also in the way it controls the asymptotic behaviour of the action at large, slowly varying fields. For positive (negative) Λ the Euclidean action is bounded (unbounded) from below for large spheres (so large that the curvature term can be neglected). This has no implication classically, since there is no stationary point of the action out there. However, at the quantum level, the situation, for negative Λ, is very much the same as that of a false vacuum [26] which, non-perturbatively, can tunnel and decay.

Does then quantum gravity make sense at all for negative CC? Does one get into a runaway situation of instability which would resemble the one we have encountered with wormholes, or do quantum effects recover the sick theory through a process of quantum resuscitation [27]? And, in this latter case, is one led to a new vacuum with a drastically different (and hopefully smaller) effective CC?

Whatever the answer, it looks to me that the abrupt, qualitative change of the theory as Λ goes through 0 – a sort of phase transition – should have something to do with the eventual resolution of one of the most out (and long)-standing challenges that Nature has left to us.

REFERENCES

[1] A. Einstein, *Sitz. Ber. Preuss. Akad. Wiss.* (1917) 142.

[2] For a nice recent review, see S. Weinberg, "The Cosmological Constant Problem", Morris Loeb Lectures in Physics, Harvard University, Univ. of Texas preprint UTTH-12-88 (1988).

[3] See, e.g., Inflationary Cosmology, Editors L. Abbott and So.Y. Pi, World Scientific Publishing Co. (1986).

[4] E. Baum, *Phys. Lett.* **133B** (1983) 185.

[5] S.W. Hawking, *Phys. Lett.* **134B** (1984) 403.

[6] S. Coleman, *Nucl. Phys.* **B310** (1988) 643.

[7] See, for instance:
G.W. Gibbons, S.W. Hawking and M.J. Perry, *Nucl. Phys* **B138** (1978) 141;
S.W. Hawking, *Nucl. Phys.* **B144** (1978) 349, **B239** (1984) 257;
J.B. Hartle and S.W. Hawking, *Phys. Rev.* **D28** (1983) 2960.

[8] S.W. Hawking, *Phys. Lett.* **135B** (1987) 337; *Phys. Rev.* **D37** (1988) 904; S.B. Giddings and A. Strominger, *Nucl. Phys.* **306** (1988) 890.

[9] I. Klebanov, L. Susskind and T. Banks, *Nucl. Phys.* **B317** (1989) 665.

[10] W. Fischler, I. Klebanov, J. Polchinski and L. Susskind,*Nucl. Phys.* **B327** (1989) 157.

[11] G. Veneziano, *Mod. Phys. Lett.* **4A** (1989) 695.

[12] W.G. Unruh, "Quantum Coherence, Wormholes, and the Cosmological Constant", Santa Barbara preprint NSF-ITP-88-168 (1988).

[13] S.W. Hawking, "Do Wormholes Fix the Constants of Nature?" DAMTP preprint (1989).

[14] G. Veneziano, Invited Talk at the Annual Meeting of the Italian Phys. Soc. (Naples, Oct. 1987);
D.J. Gross, Proc. XXIVth Int. Conf. on High Energy Physics, (Munich, Aug. 1988, R. Kotthaus and J.H. Kuhn Eds., Springer Verlag Publ. Co.) p. 310.

[15] D. Amati, M. Ciafaloni and G. Veneziano, *Phys. Lett.* **B216** (1989) 41;
G. Veneziano, Proc. of Superstring'89 Workshop, Texas A& M University, R. Arnowitt et al. Eds., World Scientific, Singapore, 1990), p. 86.

[16] T.R. Taylor and G. Veneziano, *Phys. Lett.* **212B** (1988) 147;
R. Brandenberger and C. Vafa, *Nucl. Phys.* **B316** (1988) 391;
J.J. Atick and E. Witten, *Nucl. Phys.* **310** (1988) 291; K. Konishi, G. Paffuti and P. Provero, *Phys. Lett.* **B234** (1990) 276.

[17] R. Petronzio and G. Veneziano, *Mod. Phys. Lett.* **A2** (1987) 707, and references therein.

[18] S. Coleman, private communication.

[19] S. Coleman and E. Weinberg, *Phys. Rev.* **D7** (1973) 1888.

[20] T.R. Taylor and G. Veneziano, *Phys. Lett.* **228B** (1988) 480.

[21] S.M. Christensen and M.J. Duff, *Nucl. Phys.* **B170** (1980) 480;
E.S. Fradkin and A.A. Tseytlin, *Nucl. Phys.* **B201** (1982) 469.

[22] T.R. Taylor and G. Veneziano, *Nucl. Phys.* **B345** (1990) 210.

[23] G. Vilkovisky, in "Quantum Theory of Gravity", S.M. Christensen Ed., (Adam Hilger, Bristol, 1984) p. 169; *Nucl. Phys.* **B234** (1984) 509;
B.S. DeWitt in "Architecture of Fundamental Interactions at Short Distances", Les Houches, Session XLIV, P. Ramond and R. Stora, Eds., (Elsevier Science Pub. Co., 1987) p. 1023.

[24] M. Kăc, *Amer. Math. Monthly* **73** (1966) 1.

[25] B.S. De Witt, *Phys. Rev.* **162** (1967) 1239.

[26] S. Coleman, The Uses of Instantons in "The Whys of Subnuclear Physics", (Erice 1977, Plenum Publishing Co., New York, 1979).

[27] S. Dimopoulos and H. Georgi, *Phys. Lett.* **B117** (1982) 287.

VELOCITY OF PROPAGATION OF GRAVITATIONAL RADIATION, MASS OF THE GRAVITON, RANGE OF THE GRAVITATIONAL FORCE, AND THE COSMOLOGICAL CONSTANT

J. Weber

University of Maryland
College Park, Maryland 20742
and
University of California
Irvine, California 92717

ABSTRACT

Einstein's equations with cosmological constant are considered. With an appropriate set of coordinates, the vacuum equations have the same form as the Klein Gordon Equation.

The range of the gravitational force is the Compton Wavelength of the graviton with rest mass m. The Cosmological Constant is one half the reciprocal of the squared Compton Wavelength.

Limits on the graviton mass are obtained by considering the observational data on the advance of the perihelion of Mercury, the observed gravitational radiation from Supernova 1987A, and the known gravitational binding of clusters of galaxies. The observed pulses from Supernova 1987A are in good agreement with the cross section theory published in 1984 and 1986, and reviewed here.

As first discussed by F. Zwicky, the graviton mass, as deduced from known gravitational binding of clusters of galaxies, is less than 1.2×10^{-63} grams. The Cosmological Constant is less than 6.4×10^{-52} cm^{-2}.

INTRODUCTION

Einstein's equations are

$$R_{\gamma\nu} - \frac{1}{2} g_{\gamma\nu} R - \lambda g_{\gamma\nu} = \frac{8\pi G}{c^4} T_{\gamma\nu} \qquad (1)$$

In (1) $R_{\gamma\nu}$ is the Ricci tensor, R is the curvature scaler, $g_{\gamma\nu}$ is the metric tensor, λ is the Cosmological Constant, $T_{\gamma\nu}$ is the stress energy tensor, G is Newton's constant of gravitation, and c is the speed of light.

WEAK FIELD SOLUTIONS

For weak fields, let

$$g_{\gamma\nu} = \delta_{\gamma\nu} + h_{\gamma\nu} \tag{2}$$

$\delta_{\gamma\nu}$ is the Lorentz metric and $h_{\gamma\nu}$ is a first order quantity. To first order, the Ricci tensor may be written as

$$R_{\gamma\nu} = -\frac{1}{2}\delta^{\sigma\alpha}h_{\gamma\nu,\sigma\alpha} - \frac{1}{2}\left[\left(\frac{1}{2}\delta_\gamma^\beta h - h_\gamma^\beta\right)_{,\beta\nu} + \left(\frac{1}{2}\delta_\nu^\beta h - h_\nu^\beta\right)_{,\beta\gamma}\right] \tag{3}$$

Coordinates are chosen such that

$$\left(h_\gamma^\beta - \frac{1}{2}\delta_\gamma^\beta h\right)_{,\beta} = 0 \tag{4}$$

In (4) h is the trace h_α^α. The Ricci tensor is then given by

$$R_{\gamma\nu} = -\frac{1}{2}\delta^{\sigma\alpha}h_{\gamma\nu,\sigma\alpha} \tag{5}$$

Einstein's equations (1) are in this approximation

$$-\frac{1}{2}\delta^{\sigma\alpha}h_{\gamma\nu,\sigma\alpha} - \frac{1}{2}g_{\gamma\nu}\left(-\frac{1}{2}\delta^{\sigma\lambda}h_{,\sigma\lambda}\right) - \lambda g_{\gamma\nu} = \frac{8\pi G}{c^4}T_{\gamma\nu} \tag{6}$$

In this order (6) may be written as

$$-\frac{1}{2}\delta^{\sigma\alpha}\left(h_{\gamma\nu} - \frac{1}{2}\delta_{\gamma\nu}h\right)_{,\sigma\alpha} - \lambda g_{\gamma\nu} = \frac{8\pi G}{c^4}T_{\gamma\nu} \tag{7}$$

New field quantities $\phi_{\gamma\nu}$ and $\phi_\gamma^\gamma = \phi$ are defined by

$$\phi_{\gamma\nu} = h_{\gamma\nu} - \frac{1}{2}\delta_{\gamma\nu}h \tag{8}$$

$$\phi_\gamma^\gamma = \phi = h - 2h = -h \tag{9}$$

$$\phi_{\gamma\nu} = g_{\gamma\nu} - \delta_{\gamma\nu} - \frac{1}{2}\delta_{\gamma\nu}h = g_{\gamma\nu} - \delta_{\gamma\nu} + \frac{1}{2}\delta_{\gamma\nu}\phi \tag{10}$$

In terms of $\phi_{\gamma\nu}$, Einstein's equations are

$$-\frac{1}{2}\Box\phi_{\gamma\nu} - \lambda g_{\gamma\nu} = \frac{8\pi G}{c^4}T_{\gamma\nu} \tag{11}$$

Raising one index and summing gives

$$-\frac{1}{2}\Box\phi - 4\lambda = \frac{8\pi G}{c^4}T \tag{12}$$

Einstein's equations are then, in first order

$$-\frac{1}{2}\Box g_{\gamma\nu} - \frac{1}{4}\delta_{\gamma\nu}\Box\phi - \lambda g_{\gamma\nu} = \frac{8\pi G}{c^4}T_{\gamma\nu} \tag{13}$$

(12) and (13) give

$$-\frac{1}{2}\Box g_{\gamma\nu} - \lambda g_{\gamma\nu} + 2\lambda\delta_{\gamma\nu} = \frac{8\pi G}{c^4}\left(T_{\gamma\nu} - \frac{1}{2}\delta_{\gamma\nu}T\right) \tag{14}$$

In first order

$$2\lambda\delta_{\gamma\nu} = 2\lambda g_{\gamma\nu} \tag{15}$$

(14) then gives

$$-\frac{1}{2}\Box g_{\gamma\nu} + \lambda g_{\gamma\nu} = \frac{8\pi G}{c^4}\left(T_{\gamma\nu} - \frac{1}{2}\delta_{\gamma\nu}T\right) \tag{16}$$

In vacuum

$$\Box g_{\gamma\nu} = 2\lambda g_{\gamma\nu} \tag{17}$$

(17) are the Einstein equations for the vacuum, with Cosmological Constant, in first approximation.

KLEIN GORDON EQUATION

In our study of Quantum Mechanics we begin with the non-relativistic Schroedinger equation, then the results are extended for the Special Theory of Relativity by writing

$$E = \sqrt{p^2c^2 + m^2c^4} \tag{18}$$

In (18) E is the energy, p is the momentum, m is the rest mass. E and p are replaced by the operators

$$E \rightarrow i\hbar\frac{\partial}{\partial t} \tag{19}$$

$$p \rightarrow -i\hbar\nabla \tag{20}$$

(18), (19), and (20) and introduction of the Dirac matrices lead to the Dirac equation. If (18) is squared and E and p replaced by (19) and (20) we have the Klein Gordon equation.

$$\Box\psi = \frac{m^2c^2}{\hbar^2}\psi \tag{21}$$

Comparing (21) and (17) suggests that

$$\lambda = \frac{m^2c^2}{2\hbar^2} \tag{22}$$

RANGE OF THE GRAVITATIONAL FORCE

Free particles without spin satisfy the geodesic equation

$$\frac{d^2x^\gamma}{ds^2} + \Gamma^\gamma{}_{\alpha\beta} \frac{dx^\alpha}{ds} \frac{dx^\beta}{ds} = 0 \tag{23}$$

For a particle at rest, (23) is for mass m

$$m \frac{d^2x^\gamma}{dt^2} = -mc^2 \Gamma^\gamma_{oo} \tag{24}$$

The gravitational force is then

$$\text{FORCE} = mc^2 g^{\gamma\nu} g_{oo,\nu}\left(\frac{1}{2}\right) \tag{25}$$

A solution of (17) is

$$g_{oo} = \frac{A e^{\frac{-mcr}{\hbar}}}{r} - 1 \tag{26}$$

(26) implies a range given by

$$\text{RANGE} = \frac{\hbar}{mc} \tag{27}$$

(27) and (22) give

$$\text{RANGE} = \sqrt{\frac{1}{2\lambda}} \tag{28}$$

Experimental data on m or the range then enable us to set limits on λ.

VELOCITY OF PROPAGATION OF GRAVITATIONAL INTERACTIONS

The momentum p is given for a particle with velocity v by

$$p = \frac{mc}{\sqrt{\left(1 - \frac{v^2}{c^2}\right)}} \tag{29}$$

(29) and (18) lead to

$$E = \frac{mc^2}{\sqrt{1 - \frac{v^2}{c^2}}} \tag{30}$$

Observational data enable us to set limits on the velocity of propagation of gravitational interactions.

Data to be discussed are the advance of the Perihelion of Mercury, and the Supernova 1987A observations.

PERIHELION ADVANCE

The advance of the Perihelion of planetary orbits is given by Δ with

$$\Delta = \frac{6\pi GM}{c_g^2 a (1-\epsilon^2)} \tag{31}$$

In (31) M is the mass of the sun, c_g is the velocity of propagation of gravitational interactions, a is the semi major axis and ϵ is the eccentricity. If c_g is the velocity of light, for the planet Mercury, Δ = 43 seconds per century.

The observed value for Δ is 42.6 ± 0.9 seconds per century. It is unlikely that the difference between observed and theoretical value is due to a velocity of propagation of gravitational interactions smaller then the speed of light. In any case, much smaller limits result from the Supernova 1987A observations.

SUPERNOVA 1987A OBSERVATIONS AND GRAVITATIONAL ANTENNA CROSS SECTIONS

Gravitational Radiation Antennas at the University of Rome and Maryland received coincident pulses during the Supernova 1987A rapid evolutionary period. These data imply velocity of propagation of gravitational interactions very close to the velocity of light.

In 1960, the theory of the absorption cross section of gravitational radiation antennas was given. The elastic solid antenna was assumed to be a continuous elastic solid with quadrupole moment associated with the total mass.

In 1984 and 1986 a new theory was given. The new theory regarded the antenna as a large number of small quadrupoles associated with the large number of atoms. The 1984, 1986 theory, published before Supernova 1987A, gave a much larger cross section than the 1960 theory. Observed pulse heights are in good agreement with the 1984, 1986 theory which is presented here.

The S matrix for interaction of gravitational radiation with a single quadrupole of reduced mass m and mass separation r^l is given by

$$S = \frac{mc}{\hbar} \int \langle F | (R_{oloj} + R^+_{oloj})(q^j + q^{j+}) r^l \, \overline{\psi}_a \psi_a | o \rangle \, d^4x \tag{32}$$

In (32) $<F|$ is the final state, $|o>$ is the original state, q^j is a harmonic oscillator coordinate operator. R_{oloj} is the second quantization Riemann tensor given by

$$R_{oloj} = R^{KM}_{oloj} \exp\left[-\left(\frac{i}{\hbar}\right) \overline{p}_k \cdot \overline{r}\right] d_k \tag{33}$$

In (33) R^{KM}_{oloj} is a normalization constant, and d_k are annihilation operators. $\overline{\psi}_A$ is a creation operator and ψ_A is an annihilation operator for a quadrupole.

For N quadrupoles in a volume with linear dimensions small compared with a wavelength, (32) is evaluated as

$$S_n = \frac{Nmc^2}{\hbar\sqrt{2\pi}} \int \sum_k R^{KM}_{oloj} \, r^1 \, q^j_{v\pm 1, v} \exp[i(\omega \pm \omega_0)t]\,dt \quad (34)$$

In (34) $q_{\pm 1, v}$ are harmonic oscillator matrix elements. (34) is evaluated as

$$S_n = \frac{Nmc^2\sqrt{2\pi}}{\hbar} R(\omega_0)_{oloj} \, r^1 \, q^j_{v\pm 1, v} \quad (35)$$

In (35) $R(\omega_0)_{oloj}$ is the Fourier transform of the Riemann tensor. (35) gives the cross section

$$\sigma_n = \frac{2\pi N^2 m^2 c^4 V \, |R(\omega_0)_{oloj} \, r^1|^2 \, (q^2_{v+1,v} - q^2_{v-1,v})}{\hbar^2 \tau c} \quad (36)$$

In (36) τ is the time, V is the volume. $R(\omega_0)_{oloj}$ may be written in terms of the cross sectional area A; for normalization of one graviton per unit volume,

$$|R_{olo}(\omega)|^2 = \frac{4\pi G \hbar \omega^2}{A c^7} \quad (37)$$

The total cross section (36) becomes

$$\sigma_n = \frac{8\pi^3 N^2 G m r^2}{c^2 \lambda} \quad (36A)$$

In (36A), λ is the gravitational wavelength. (36A) does not include heat bath interactions. Both classical and quantum theory give for the absorbed energy in the presence of heat bath interactions

$$U = \frac{D}{2\pi} \iiint \frac{\omega \omega' c^4 R^\gamma_{o\alpha o}(\omega) \, r^\alpha \, R^\gamma_{o\beta o} \, r^\beta \, e^{i(\omega - \omega')t} \, d\omega \, d\omega' \, dt}{(-\omega^2 + i\omega D/m + k/m)(-(\omega')^2 - i\omega' D + k/m)} \quad (38)$$

(38) gives the modified cross section

$$\sigma_n = \frac{8\pi^3 M L^3 Q_1 G}{12 c^2 \lambda S_a} \quad (39)$$

In (39) M is the total mass, L is the length, Q_1 is the "quality factor" of a single quadruple, S_a is the length occupied by one atom. For the antennas operating during SN 1987A

$$\sigma_n \to 2 \times 10^{-18} \text{ cm}^2 \quad (40)$$

EXPERIMENTAL DATA

Correlations were observed with neutrino detectors at Mont Blanc, Kamioka, Baksan, and the Irvine Michigan Brookhaven detector in Ohio, the Frejus Muon detector and the Monte Rosa Cosmic ray detector.

In all cases the gravitational pulses arrived first, delays varied between about 1.0 and 3.0 seconds. If we assume an arrival time uncertainty of one second, transit time 165000 years = 5.2 x 10^{12} seconds

$$\frac{v}{c} = 1 - 2 \times 10^{-13} \tag{41}$$

For angular frequency $\omega \sim 10^4$, $E \sim 10^{-23}$ ergs

$$E = \frac{m_0 c^2}{\sqrt{1 - \frac{v^2}{c^2}}} \quad \text{gives}$$

$$m_0 < 7 \times 10^{-51} \text{ grams} \tag{42}$$

$$\frac{\hbar}{m_0 c} = 4.8 \times 10^{12} \text{ cms} \tag{43}$$

$$\lambda < 2.2 \times 10^{-26} \text{ cm}^{-2} \tag{44}$$

Smaller limits follow from a much earlier analysis of F. Zwicky who believed that clusters of galaxies did not form super clusters. That would imply

$$\text{Gravitational Range} \approx 2.8 \times 10^{25} \text{ cms}$$

$$m < 1.2 \times 10^{-63} \text{ grams} \tag{45}$$

$$\lambda < 6.4 \times 10^{-52} \text{ cm}^{-2} \tag{46}$$

It is currently believed that super clusters do form and the limits are in fact smaller than (45) and (46).

CONCLUSION

The smallest limits on the cosmological constant and graviton mass result from considering the fact that the gravitational force range is larger than dimensions of galactic clusters.

REFERENCES

1. J. Weber, General Relativity and Gravitational Waves -- Interscience, Wiley 1961, and the University Xerox Microfilms, Foundations of Physics. Volume 14, 12, pages 1185-1209, 1984, Volume 3 Eddington Centenary Symposium, World Scientific 1986 pages 1-77 Edited by T.M. Karade and J. Weber.

2. Aglietta et al Il Nuovo Cimento 12C., Gennaio Febbraio 1989 page 75.

CONTRIBUTORS

Stephen L. Adler
Institute for Advanced Study
Princeton, NJ 08540
USA

Peter G. Bergmann
Departments of Physics
Syracuse University , Syracuse,   NY 13244-1130
New York University , New York, NY 10003
USA

Nicola Dallaporta
Dipartimento di Astronomia, Università di Padova
Vicolo Osservatorio, 535623   Padova
Italia

Venzo de Sabbata
Dipartimento di Fisica, Università di Bologna
Via Irnerio ,46
I-40126 Bologna
Italia

S.B. Fadeev
USSR State Commitee for Standards
9 Leninski Prospect, 117049, Moscow
USSR

Fang Li Zhi
Institute for Advanced Study
Princeton, New Jersey 08540
USA

J.N. Goldberg
Department of Physics, 201 Physics Building
Syracuse University ,Syracuse, NY 13244-1130
USA

V.D. Ivashchuk
USSR State Commitee for Standards
9 Leninski Prospect, 117049, Moscow
USSR

Mark Israelit
School of Education of the Kibbutz Movement
University of Haifa, Oranin, Tivon
Israel

I.M. Khalatnikov
L.D. Landau Institute for Theoretical Physics,
USSR Academy of Sciences
GSP-1 117940 ul. Kosygina 2, Moscow V-334
USSR

S. Lauro
Department of Physics
St. John's University, Jamaica,
New York , N.Y. 11451
USA

G. Lavrelashvili
Tbilisi Mathematical Institute
380093 Tbilisi
USSR

A. Linde
Stanford University
Department of Physics, Varian Bldg
Stanford, California 94305-4060
USA

Vladimir S. Manko
Department of Theoretical Physics
Peoples' Friendship University
Moscow U.S.S.R.

V.N. Melnikov
USSR Committee for Standards
9 Leninski Prospect, 117049 Moscow
USSR

Igor D. Novikov
Astro Space Centre of P.N. Lebedev Physical Institute
of the USSR Academy of Sciences
Profsoyuznaja 84/32, 117810 Moscow
USSR

Roger Penrose
Mathematical Institute
University of Oxford
24-29 St Giles, Oxford OX1 3LB
United Kingdom

Guido Pizzella
Department of Physics
Università di Roma "La Sapienza", INFN-Roma
Piazzale Aldo Moro 2, 00185 Roma
Italia

D.C. Robinson
Department of Mathematics, King's College
Strand, London WC2R 2LS
United Kingdom

Nathan Rosen
Department of Physics
Technion-Israel Institute of Technology
Haifa 32000
Israel

V.A. Rubakov
Institute for Nuclear Research
Academy of Sciences of the USSR
60th October Anniversary Prospect 7a
Moscow 117 312
USSR

N. Sánchez
Observatoire de Paris
Section de Meudon, Demirm
92195 Meudon Principal Cedex
France

Ernst Schmutzer
Theoretical-Physics Institute
Friedrich Schiller Universitä
Max Wien Pl.1, 6900 Jena
Germany

E.L. Schucking
Department of Physics
New York University
New York, N.Y. 10003
USA

C. Sivaram
Dipartimento di Fisica
Università di Bologna
Via Irnerio, 46, I-40126 Bologna
Italia

C. Soterio
Department of Mathematics
King's College
Strand, London WC2R 2LS
United Kingdom

P.G. Tinyakov
Institute for Nuclear Research
Academy of Sciences of the USSR
60th October Anniversary Prospect 7a
117 312 Moscow
USSR

G. Veneziano
Theory Division, CERN
CH 1211, Geneva 23
Switzerland

J.-Z. Wang
Department of Physics
New York University,
New York, N.Y. 10003
USA

J. Weber
Physics Department
University of Maryland
College Park, MD 20742
USA

INDEX

Anthropic principle, 23, 98, 101, 108, 115

Baby universes, 87, 101, 104
Binaries, 17
Black holes, 11, 12, 14, 16, 163
Bogoliubov transform, 164, 165, 168, 177
Boltzmann distribution, 148

Canonical formalism, 61
Chaotic inflationary scenario, 108
Clustering, 55, 56
Cluster of galaxies, 219
Conservation laws, 183
Constant spinors, 133
Constraints, 62
Correlation function, 54
Cosmic strings, 174
Cosmological constant, 5, 19, 87, 88, 95, 101, 179, 183-185, 217
Cosmological models, 65, 66, 151
Cosmological singularity, 77, 151, 189, 195
Coupling constants, 108
Creation of universes, 88, 112
Critical density, 22
Crystal dislocations, 30

De Sitter space, 112, 168, 207
Dirac brackets, 59
Dissipation free regime, 65

Distribution function, 54
Doctorate degree, 2
Double universe model, 102
Dynamical System, 65

Early universe, 27, 153, 154
Effective action, 5
Einstein, 3
Energy momentum tensor, 21
Euclidean functionnal integral, 87, 205
Exact solutions, 37
Explorer experiment, 145

Finite theory, 159
Flatness problem, 151

General relativity, 4
Grand unified theories, 102
Gravitational antenna cross section, 221
Gravitational collapse, 11
Gravitational waves, 139, 149, 217, 220

Hawking radiation, 157
Helicity, 124
Hubble constant, 21

Inflation, 101, 108, 202
Instantons, 87
Isometric embedding, 196

227

Kähler metrics, 190
Kaluza-Klein theories, 102
Klein-Gordon equation, 219

Laplace transform, 5
Large scale structure, 51
Laws of motion, 2
Lyman-Alpha absorption lines, 51

Magnetic dipole moment, 121
Magnetostatic solution, 122
Mass shedding, 15, 16
Matter-dominated period, 156
Mellin transform, 5
Minisuperspace model, 88
Möbius strip, 197
Multidimensional gravity, 37

Naked singularities, 11, 18
Neutron stars, 11, 12, 16
Non-linear graviton construction, 129, 136
Non-locality, 213
Non-resonant detectors, 141
Null surface, 59

Observables, 3

Particle transmutation, 165
Pencil-beam surveys, 57
Perihelion advance, 221
Phase space, 65
Planck energy density, 102
Planck epoch, 21
Planck length, 87, 88, 130
Planck mass, 5, 158
Projective space, 189
Projective unified field theory, 179
Pulsars, 11, 13

Quadrupole, 16
Qualitative methods, 65
Quantum cosmology, 101
Quantum effects, 19

Quantum fluctuations, 112
Quantum gravity, 6, 157, 201
Quantum state, 3

Radiation-dominated period, 156
Regular differential equations, 65
Renormalization group, 158
Resonant detectors, 142
Ricci-flat manifolds, 124

Scattering amplitude, 176
Self-dual solutions, 129-136
Shock waves, 166
Space-time singularities, 179
Space-time topology, 6
Spin-density tensor, 24
Spin 3/2 charges, 124
Stationary electrovacuum solution, 121, 124
Stellar evolution, 6
String theory, 157
String instabilities, 169
Supernova 1987 A, 221
Superstrings, 165
Supersymmetry, 23
Symmetry breaking, 21

Torsion, 19
Third quantization, 87, 88
Topological changes, 87
Topological effects, 201
Twistors, 124, 132, 136
Thermodynamics, 154
Theory of every thing, 159
Tolman's cosmos, 189, 197

Vaccum energy, 21, 23, 102
Vacuum space-times, 124
Velocity propagation, 220
Viscosity, 65, 76, 77

Wave function of the universe, 88
Wormholes, 6, 24, 87, 106, 201